T0107178

‘Erudite, funny and sad – a glorious roller coaster of a book whose twists and turns take us again and again to the dissolving edges between reality and mirage’

Jean Sprackland, author of *These Silent Mansions*

‘Reading *Strandings* is to be caught in a magical drift, borne ever deeper, into the atavistic, into the animal selves, still alive, inside us. I was captivated’

David Keenan, author of *This Is Memorial Device*

‘A funny, personal and poetic dive into the mystical world of whale strandings. A compelling and fascinating read’

Will Sharpe, creator of Channel 4’s *Flowers*

‘I devoured this – it’s wonderful. As compellingly eccentric as it is deeply humane, emotionally and politically astute. One test of a memoir is that you actually want to spend time with the narrator, and Riley is so charming, outward looking and rigorously honest it’s impossible to consider anyone not. *Strandings* is a funny and poignant exploration of a fringe I didn’t know existed, but written with such engaging personality and prose style I feel as though I’ve met the cast myself’

Luke Kennard, author of *The Transition*

‘Just the kind of book we need now: ecologically, politically, comedically potent and with personality worthy of Melville’

Liam Williams, comedian and writer, BBC3’s *Pls Like*

STRANDINGS

STRANDINGS

Confessions of a Whale Scavenger

PETER RILEY

P

PROFILE BOOKS

First published in Great Britain in 2022 by
Profile Books Ltd
29 Cloth Fair
London
EC1A 7JQ

www.profilebooks.com

1 3 5 7 9 10 8 6 4 2

Typeset in Sabon by MacGuru Ltd
Printed and bound in Great Britain by Clays Ltd, Elcograf S.p.A.

A CIP catalogue record for this book is available from the British Library.

ISBN 978 1 78816 607 2
eISBN 978 1 78283 749 7

For my mum, Claudia Riley, who loved me

STRANDINGS

Contents

1

Hunstanton

give him line and scope,
Till that his passions, like a whale on ground,
Confound themselves with working.

William Shakespeare, *Henry IV*, Part II, 4.4 (1597)

When I was thirteen, I helped a woman with blue hair load the jaw of a sperm whale into the back of a yellow Volvo 245. It only just fitted. What she'd got hold of wasn't quite as big as the one that greets you at the entrance of the Natural History Museum in Oxford; that's still the most enormous jaw I've ever seen. Nevertheless, what I helped carry was big. And heavy. Add to that the pounds of blubber and you get a sense of what we transported that morning – maybe the weight of a tall man. According to the butchers I've asked, it must have taken her at least half an hour to saw through. If you've ever handled a piece of whalebone, you'll know how durable and solid it feels – like reinforced, triple-weighted pumice. In the case of a sperm whale, it's even sturdier, needing to withstand higher water pressures than in other, shallower-diving members of its species. The blue-haired woman had accomplished this at night, alone, and in the steady Norfolk rain.

The sperm whale in question had stranded itself the day before on the iliac crest or hip of eastern England. Old Hunstanton's north beach is a stretch of sand and mud that slides into the Wash, one of Britain's largest estuaries. It was 1997, the first weekend of the Easter break, and my mum had rented a small barn just outside Holme-next-the-Sea, a nearby village that would, a year later, become known for its 'Seahenge' – a 4,000-year-old Bronze Age timber circle of fifty-five posts surrounding an upturned oak stump sunk into the sand.

There was a break in the rain, and the very early morning horizon already shone through a light blanket of coastal cloud. In a neighbouring field, the hares were beginning their daily boxing and sprinting. I had written 'Gone to visit the whale' in my neatest handwriting and left the note on the doormat.

We'd driven up from the south-west suburbs of London in a light-blue Ford Fiesta. Mum and I had recently moved into a place that stood about twenty metres from a stretch of the Windsor-to-Waterloo line. Four trains an hour. In spite of everything, it all ran fairly smoothly. I was a member of the Fifth Staines Sea Scouts, which meant that once a week I swore allegiance to the Queen and the Union Jack. I was shy and polite. My dad would dutifully pick me up on Saturday mornings and take me clockwise past Heathrow and up the M25 to see his mother in Watford. We'd eat cream of mushroom soup from a can, and everyone would say how delicious it was. My mum held down a series of jobs. She saved up enough to put herself through university – I was nine when I went to her graduation. She began working in adult education, teaching German. On Tuesday and Wednesday evenings she taught language classes in our front room to groups of adults – Tony, Robert, Reg and Sheila. They all greeted me whenever I passed through the living room on my way to the toilet. Robert once took Mum and me to the South Coast and Lyme Regis, where the fossil remains of 180-million-year-old marine reptiles regularly wash out of the cliffs.

The afternoon before I helped load the sperm whale's jaw, I had joined a crowd of maybe fifty people that had converged on the animal. As is usually the case, one or two tried to think of ways to float it back out to sea. A man wearing white chinos had taken it upon himself to sprint back and forth between whale and surf armed only with his toddler daughter's crab bucket. The child ran after him angrily demanding its return. The rest of us stood and watched as he tossed small amounts of water onto the whale's head. After maybe twenty buckets, a fluorescent-jacketed man who was putting up a cordon of red and white tape informed us all that if the creature was not already dead, then it would very likely soon be. The best and kindest thing to do, he urged us, would be to leave it in peace. 'A whale of this size', he said, 'once stranded, tends to suffocate very quickly under its own weight.' The father received a smattering of applause as he re-joined the assembled crowd. His daughter continued to cry.

The next morning, the whale's jaw appeared over the crest of a beachgrass dune, followed by a face concentrated in the effort to keep it balanced as the bowsprit of a wheelbarrow. The blue-haired woman had wrapped her prize in a white sheet that fluttered bloodily in the breeze. I knew immediately what she'd done. A butcher's bowsaw sheathed under her arm, she wore a pair of industrial-looking rubber gloves that were now smeared in a thick abattoir slime. For a few moments she carried on struggling, unaware of the child standing just a few metres away. At that moment she was easily the most beautiful person I'd ever seen. Her blue hair was cropped short. The bridge of her nose was tanned and freckled. I stood dead still. She stopped; looked at me – inhaled a lungful of sea air. Setting the wheelbarrow down, she transferred the saw to her right hand, drew herself up a good foot taller and faced me in silence. It occurred to me that I was about to be murdered. A minute passed before she spoke.

I now know that she was part of a wide circle of cetacean body-snatchers and bone-collectors. When the whales appear, the scavengers move in. There are more than you might think. Take this hooded figure, snapped for the *Daily Mail* in late 2011. Notice that the jaw's already gone. The accompanying caption explained, with some restraint: 'Member of the public cuts off a tooth from the beached whale, which washed up on the Norfolk coast on Christmas Eve.'*

By early January, a follow-up story described how local police had detained a local 'male youth' for subsequently

* Charles Walford, 'Monster from the deep ... on the Norfolk coast: 40ft sperm whale washes up on Christmas Eve.' *Daily Mail*, 28 December 2011.

posting a bill of sale on a social networking site; £5 per individual tooth (fifteen available) or £45 the lot (fifteen teeth plus eleven still in situ, plus the jaw itself). How had he settled on the going rate for a whale in 2011? There are a few considerations he might have weighed up. First, under International Whaling Commission regulations there's a ban on 'harvesting' (though harvesting really only applies to whaling proper), as well as a ban on 'trade', under the Convention on International Trade in Endangered Species (that's one he might have been cautioned for). Then there's the Conservation of Habitats and Species Regulations Act of 2010, which promises up to six months in prison and an unlimited fine for this kind of activity. Contraband once, twice and maybe three times over (without a specific buyer in mind), the youth's modest pricing reflected some challenging market conditions.*

* It is rare that prosecutions take place, but sometimes they do. In 2012 a Fife man, Steven Paterson, was sentenced to 160 hours community

Amateur photographer and local whale vigilante Daryl Hind captured this next scene. This is Skegness in Lincolnshire, and one of thirty sperm whales stranded at the beginning of 2016 (one of the largest mass strandings of *Physeter macrocephalus* on record). Daryl sold his picture to the *Daily Express*. The perpetrator is clearly an opportunist rather than a member of the inner circle. You need more than a kitchen knife and a pair of pliers to extract the tooth of a sperm whale.

These newspaper accounts interest me because the journalists who write them often don't quite know what it is they're describing, or how they ought to respond.

service for trading in 'dead endangered animal parts'. Paterson was caught when the UK Border Agency discovered his website offering sperm whale teeth to foreign buyers. Having solved the case, PC Ian Laing, Fife Constabulary's Wildlife and Crime Co-ordinator, assured the public that he would 'leave no stone unturned in reducing wildlife crime'.

Consequently the reports tend to veer in strange directions. The *Mail*, for example, isn't sure whether it's describing a form of criminal or at least antisocial behaviour – or whether it should in fact be praising a demonstration of entrepreneurial spirit during hard times. In the *Express*, the whale mushroomed into a '30,000-tonne mammal'.* To be clear, 30,000 tonnes is equivalent to 300 fully laden Boeing 747s; a bull sperm whale weighs in at about 30 tonnes, or two London buses.

When the blue-haired woman spoke her voice was deeper than I expected. If I wouldn't mind holding it steady. Not much further now.

Balancing it mid-transit, my hands and woollen sleeves covered in slime, she told me to watch out where I was treading. I looked down too late and half twisted my ankle on the remains of a fire pit. The jaw sagged to one side, nearly toppling us over. We regained our balance and came to a halt. It was heavy for me, even though she was doing most of the work.

'You all right?'

'Yes.' I said this without looking at her.

I had only a vague sense of the crime I was aiding and abetting. I knew that sperm whales, like sturgeon fish and swans, belonged to the Queen, and I hoped she wouldn't miss this bit. Fairly common knowledge this: in the UK, stranded whales (and sturgeon) are 'Royal Fish'. At first it

* 'PICTURED: Sick trophy hunter rips teeth from dead sperm whale washed up on UK beach: A HEARTLESS trophy hunter was caught ripping TEETH from one of the dead sperm whales which washed up on a British beach.' *Daily Express*, 27 January 2016.

might seem like a charming honorary designation. It's not. It's an assertion of private property, an oceanic extension of Enclosure and the destruction of the Commons. In his *Commentaries on the Laws of England* (1765–70), the Vinerian Professor of Law at Oxford and Tory MP William Blackstone argued that the 'Royal Prerogative', first put on the books by Edward II in 1324, was 'grounded on the consideration of his guarding and protecting the seas and coasts from pirates and robbers' – or from anyone who might wish to claim a carcass or part of a carcass for themselves.* A royal protection racket.

Steadying the jaw was only possible, I found, by hanging on to a couple of tea-coloured teeth (one in each hand), which were not so slippery when grasped through the sheet. It took us about five minutes to reach the car, which was parked on a dirt track that ran parallel to the dunes. 'Hold it steady,' she told me, an increasing urgency in her voice, as she set the barrow down and ran round to the passenger's seat.

As she leaned in, she revealed the small of her back. The tattoo must have been a recent addition, because the comet had only been visible for a few months: Hale-Bopp. It was then at its 2,600-year perihelion. I'd been watching it hang there for weeks, about the diameter of the Moon, and its tail now streaked away from the top of her sacrum to the tip of her pelvis bone.

---

* According to the office of the Receiver of Wreck, whose responsibility it now is to deal with such Fish and who I telephoned in late 2018, the current monarch has devolved this lasting privilege to the Institute of Zoology and the Natural History Museum for research purposes. But these animals are still, by default, Her Majesty's.

In retrospect, it wasn't a particularly unusual choice for a strandings enthusiast. As the anonymous author of a 1677 pamphlet entitled 'Strange news from the deep being a full account of a large prodigious whale, lately taken in the river Wivner, within six miles of Colchester' points out, whales are 'brought to Land by the same means, and for the same Reason, as Comets are placed in the Skie, viz. as a certain sign of an insuing Judgment to fall upon that Nation over which they depend'. The tattoo was part of ancient strandings lore, an inscription of a portentous association that stretches back hundreds of years.*

* See also the pamphlet 'Wonders from the deep, or, A true and exact

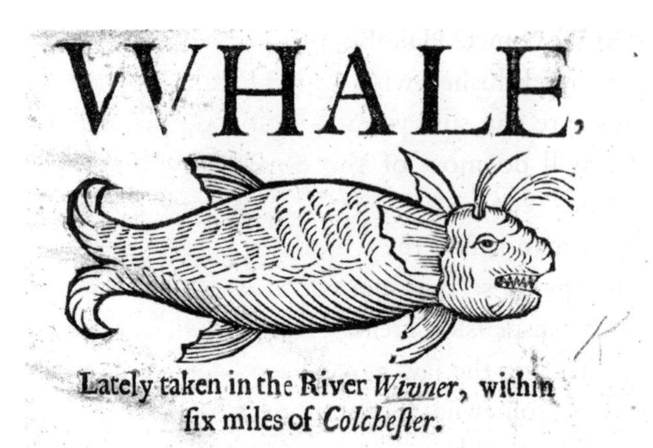

account and description of the monstrous whale lately taken near Colchester, being two and forty foot in length, and of bigness proportionable, with the manner of its coming, and being kill'd on Thursday the 9th of April, being so rare and strange a sight that multitudes of people from all parts dayly go to see it as thick as to a market or fair', London: Printed for E.W., 1677. Predictably, most of these reports and auguries at this time were written by men, but here's one by a woman: this is Sarah Bradmore's only public foray into print (who knows what else she might have written). It's a bold satire on the way Restoration England interpreted the arrival of whales: 'Strange and Wonderful have been the Effects of such unusual Creatures as Whales approaching in Rivers; so know we hath been the appearance of Comets; But I am certain, none knows what will be the Issue of these Signs and Wonders, but the Almighty who is the Occasion of them; yet some Men have had the Impudence to pretend (and put it publickly in Print) that they know the Will and Pleasure of the Almighty.' She then takes direct aim at physicians: 'According to the Constellations of the Stars which I converse with, there will happen (Anno. 1687.) a Great Rot amongst the Quack Doctors.' Why, in spite of her scepticism, did she decide to write this anti-prophecy prophecy? 'I will assure you it's very hard times, and I wanted Money.' 'So taking my leave till the coming up of the next Whale', she concludes, covering her arse, 'I remain a True and Loyal Subject to His Majesty, Whom God Preserve, and in General, your Humble Servant; S. B.', London, printed by S.J; 1686.

'Is that the comet? Hale-Bopp?'

She emerged flushed with the preparation of her cargo hold. 'Yes. Are you sure you're OK to help me lift this in? The barrow'll do most of the work – just make sure it doesn't slide.'

I helped manoeuvre the wheelbarrow into position. We tipped the jaw up so that most of its weight now rested on the car. The suspension creaked audibly. On her count, we set about heaving the jaw as far back as it would go – the process eased somewhat by the lubricating gore. In advance of the final push, it became apparent that the boot wasn't going to shut properly – in fact, a good foot of Royal Fish would have to jut out of the back of her car as she drove home, or wherever she was headed. She secured the now completely red sheet as best she could, and I helped wedge the wheelbarrow and bowsaw in on top. Then she secured the whole pile with bungee cords and finally lowered the boot until it came to rest on the jaw. Two more cords to make sure it wouldn't fly open on the drive home.

'What are you going to do with it?'

'Wash your hands in the sea,' she said, removing her gloves.

'Are you going to keep it in your garden?'

She leant down and gave me a careful hug. 'Bless you.' She released me, looked me in the eyes.

'Any time.' I meant it. I would gladly and gratefully have transported other pieces of the whale for her. Or walrus, or seal for that matter.

As she got into the driver's seat, she glanced at me one last time. I looked away. I didn't think to ask her name. She turned the engine on, inched forward to see if the jaw would hold firm, and moved slowly off down the track. She'd be in her early forties now. The blue-haired comet woman driving the yellow Volvo 245 with the jaw of a sperm whale in the boot.

In the distance early-morning dogs barked. A trail of slime traced our progress from the dunes to where I now stood alone. I sprinted down the track fifty metres or so and ducked back into the dunes. I dug a small hole in the sand, took off my irretrievably soiled woollen sweater and buried it, knowing on my return my mum would immediately ask where it was. Sheltered in the lee of the sand dune, I got the rising smell of the gore that covered my hands and wrists. Perhaps if I kept my distance from anyone I came upon, I'd just look like any other curious person come to visit the whale. It occurred to me that I might have accidentally smeared blood across my face, and I checked with a wipe of my shirt. Nothing. She would have warned me. I kept low through the dunes, and emerged onto the beach.

It lay there like a fallen moon, the aftermath of some cosmic battle. The absolute certainty of its gigantic land-turned back concealing a night's work – as though the whale had been in on it too. The strange lightness expanding across my chest as if in counterweight levitation. The binding together of a constellation: comet, whale, jaw – its beautiful scavenger. At the water's edge I doused my hands as instructed. It was like trying to wash off bacon grease with freezing water and no soap.

Fifteen years later, a return to where it began. Just along the coast at Hunstanton. Britain readying itself for the Olympic Games. A discernible swelling of national pride. I'd heard the news over the radio and immediately cancelled a class I was about to teach. It happens time and again: when a whale appears, I drop whatever I'm doing and join the others in trying to get to the body before the various well-meaning scientific and charitable organisations pull on their fluorescent vests and cordon everything off.

According to the Cetacean Strandings Investigation Programme (CSIP), there are approximately 800 strandings a year along the British coastline. Scientists still don't know for certain what keeps them coming in such numbers, but it's likely to be a combination of the increase in sonar-scrambling low-frequency noise, the warming and ever filthier oceans – and simply old age. Another recent hypothesis is cosmic in scale. As the sun releases a flare of a billion atomic bombs, the earth's electromagnetic field registers a slight modulation, tripping a cetacean switch that was navigating a whale past, say, the Uists and Benbecula in search of the northern squid that populate the Rockall Trough. The animal suddenly mistakes a due north for a west. Onboard navigation system compromised, the northern coast of Scotland segues into the soporific temperatures of the ancient sunken topography of Doggerland – the region of shallows that, 10,000 years ago, provided a land bridge between Great Britain and continental Europe. Once surrounded by the maritime treacheries of Dogger, German Bight, Forties, Humber, Fisher and Tyne, it's not long before the disorientations of the tides take over.

Look at the crowds of dead-whale watchers.*

* The owner of the local cliff-top car park took in a total of £4,022 that year, up on the previous year's takings by £3,000. 'We need a whale to come every year,' she told the local press.

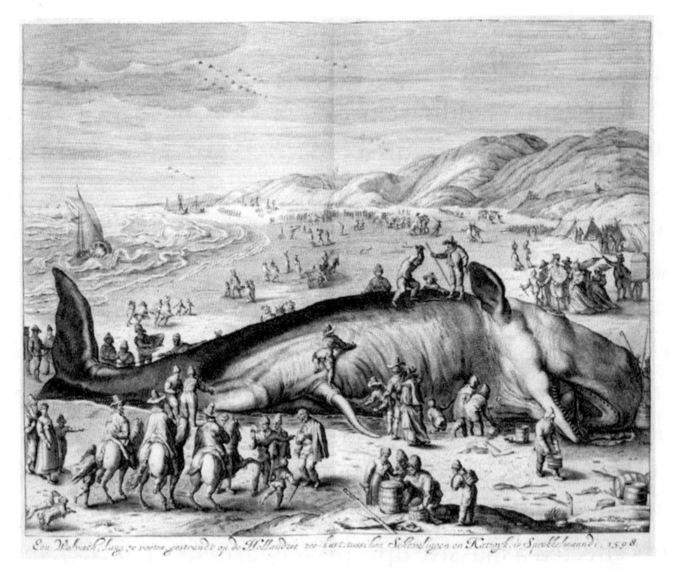

Gilliam van der Gouwen (1657–1716). Engraving is made after a draft by Hendrik Goltzius and after an engraving by Jakob Matham from the year 1598. *Copper Engraving Een Walvisch. Lang 70 voeten, gestrandt op de Hollandtse zee-kust, tusschen Scheveningen en Katwijk, in Sprokkelmaandt, 1598 in the Sylter Heimatmuseum in Keitum.*

A group of boys point at the whale's giant slumped penis. 'What is that, sir?' they ask the official on duty. They make beached whale jokes about one of their friends, and dare each other to get as close as possible. One kid, no more than twelve, reaches out a hand and slaps a fin. 'High five!' The rest of the gang explode with laughter.

FYI: the first beached whale joke in the English language occurs in Act 2, Scene 1 of Shakespeare's *The Merry Wives of Windsor* (1602). It's made at the expense of a man. Mistress Ford greets the hung-over Falstaff with: 'What tempest,

I trow, threw this whale, with so many tuns of oil in his belly, ashore at Windsor?'

Three girls standing close by, slightly older than the boys. One of them wears dream-catcher earrings. One wears fishnet tights; her fingernails are painted black. Another takes out a bottle. They pass it around, each taking healthy swigs. One raises a toast to the whale.

A young man, bottle in hand, stares directly into the whale's wide-open jaws. Seaside towns were among those hardest hit by Tory austerity policies, that great dismantling of public services that saw the net incomes of the poorest tenth of the UK population plummet between 2010 and 2015 by over a third, while bailouts, deregulation and top-end tax cuts effected a massive upward redistribution of wealth. The whale stranding as looming metaphor: a people struggling to stay afloat, washed up on the beach by yet another financial ebb tide.

An older man stands much closer than you might imagine, as though in familiar company. An alien visitor from the deep, a whale simultaneously arrives on shore as an old acquaintance, its appearance on a beach a malfunction in the natural order, evidence of man's errant stewardship of the earth – but also a homecoming of the long-lost. Whales once stalked the earth with us – as us: our shared mammalian bone structure forming on terra firma 50 million years ago. The Ngātiwai of northern New Zealand know this, and believe that whales beach themselves when they are ready to return to the fold. A leader of this Māori community will walk up to a stranded whale and welcome it back as family.

We're in Britain, though, and the consonance between stranded whale and human is different in these parts. We're looking at a non-human relative that provided the light we once wrote and worked and thought and loved by. The source of glues, soaps, face-creams, margarines, corsets and perfumes – the lube that greased and skidded capitalism into its current all-consuming delirium. Given the breathtaking extent of the harvest, can the whale even be counted as a seafaring species any more? The ones left in the sea feel like the outliers. One and all the products of dead whales, the dead whale irrevocably stranded in all of us.

And then there is this other figure perched on top of a beachgrass dune. A panoptic silhouette against the sea-bright sky. Green waders, a pair of binoculars hanging from his neck. Carrying a spade and a heavy backpack, he's on a recce. A scavenger.

What part of this animal has he come for? The jaw? Does he want to carry home a flipper? A piece of the animal's skin?* Is it for his own collection, or is he on someone's payroll? Whatever his reason for coming, I know he's part of the inner circle, those who insist on taking their fascination to another level. Sure as the tide, the jaw's gone by morning.

Nothing new. In Britain, the amateur tradition of scavenging the body parts of stranded whales stretches back centuries. In 1658, the week Oliver Cromwell died (the death knell for the Parliamentarians' short-lived republic), a boy from Blackwall spotted a 60-foot whale swimming around

---

* In 1934, twenty pilot whales washed ashore on a Gower Peninsula beach. According to the newspapers, 'souvenir hunters among the crowds of sightseers [had] been busy, for an inspection of the carcasses showed that many had had large pieces of skin cut from their backs.' See 'Rush to See 20 Monsters. Hundreds Trek to Coast,' *Daily Mail*, 9 May 1934.

the Greenwich bend of the Thames. According to one Cromwell biographer, this was 'one leviathan coming to pay his respects to another'. The poet John Dryden confirmed the connection, writing that

> first the Ocean as a tribute sent
> That Gyant Prince of all her watery Heard;
> And th'Isle when her Protecting Genius went
> Upon his Obsequies loud sighs confer'd.

As soon as this whale was mercilessly dispatched by an impromptu armada, the human scavengers moved in.

> Some bought peeces as big as a mans middle, and some took lesser peeces to shew to their neighbors, friends and acquaintance, that what is reported concerning this hugeous whaile, is an absolute truth; and as a monument of Remembrance, they do both safely and securely lock it up, esteeming more rarely of it, then a dish of Anchovis, Salmon, or Lobsters, that is a present for a Lady, for although a whale be not good to eate, it is novelty, and very strange and much more stranger to be catcht in the River of Thames so neere to London bridge.

I quote this from *Londons wonder: being a most true and positive relation of the taking and killing of a great whale neer to Greenwich; the said whale being fifty eight foot in length, twelve foot high, fourteen foot broad, and two foot between the eyes. At whose death was used harping-irons, spits, swords, guns, bills, axes, and hatchets, and all kind of sharp instruments to kill her: and at last two anchors being struck fast into her body, she could not remoove them, but the blood gusht out of her body, as the water does out of a pump. The report of which whale hath caused many hundred of people both by land and water to go and see her; the said whale being slaine hard by Greenwich upon the third day of Iune this present yere 1658.*

It is perhaps a little too comfortable for a scientifically and morally enlightened present to stare back at a particular historical moment in horror; reduce it to the perceived backwardness and barbarity of an earlier age.* Of course, I'm as susceptible to this attitude as anyone else, and agree it was a terrible shame that this beautiful animal was 'slaine hard by Greenwich'. Yet I also hold to the view that we ought in some way to attend sympathetically to what these people were doing. That it may in fact behove us to acknowledge our own curious position in relation to this lynching; trained to loathe violence against particular animals while complicit in the greatest mass extinction event since the Cretaceous–Paleogene comet struck the earth roughly 66 million years ago. 'Monuments of Remembrance', 'a present for a Lady', a 'novelty' worthy of being safely and securely locked up.

* Those early representatives of the scientific enlightenment were also there, detachedly watching and recording. The killing took place in front of the residence of committed Royalist John Evelyn (family fortune derived from gunpowder manufacture), who saved his neck during the Civil War by sticking strictly to scientific inquiry and observation. 'A large whale was taken between my land abutting on the Thames and Greenwich, which drew an infinite concourse to see it, by water, horse, coach, and on foot, from London, and all parts,' John Evelyn, diary entry, 3 June 1658.

No doubt some were on the make – indeed, some thoroughly enjoyed the spectacle of cruelty – but some carried home with them particular confluences of affect and feeling. Relics of the English Republic, souvenirs, esteemed dishes, the spoils of conflict, gifts, memento mori, runic tokens.* The Great Whale of Greenwich had an afterlife; became part of the furniture, its bones framing the entrance of a Dagenham manor house here, forming the skeleton of a Chadwell Heath High Road tollhouse there.† One of its ribs was erected as a monument on the southern boundary of

* The assumed connection between events – stranded Lord Protector/ stranded whale – persisted throughout the seventeenth and eighteenth centuries. When two further whales appeared near Gravesend a couple of decades later in 1680, another pamphlet noted that '[S]uch accidents never happen but they are presently followed by the Death or Fall of some great person or other: as there happened a like accident with this of the coming up of a great Whale into the Thames not long before the death of that much greater Monster, the Usurper Cromwell.' See 'Strange news from Gravesend and Greenwich being an exact and more full relation of two miraculous and monstrous fishes first discovered in Rainham Creek, and afterwards pursued by fishermen up the river of Thames, who with harping irons and fish-spears kill'd the biggest of them at Gravesend, which after thousands of people had view'd it, they hew'd in pieces and boyl'd in cauldrons for the oyl.' London: printed for J. Clarke at the Bible and Harp in Smithfield, 1680.

† Rib bones (and perhaps jaw?) were turned into a tollhouse at the junction of Chadwell Heath High Road and Whalebone Lane, north-east London. The tollhouse was destroyed during Second World War bombings. Whalebone Lane is three miles long. See also Whalebone Grove. Locals refer to the junction as Moby-Dick. Bones once framed the entrance of the Valence House Museum, a medieval manor house in the London Borough of Barking and Dagenham. They're now in the museum's Whalebone Gallery.

Hainault Forest (identified by Daniel Defoe in his 1722 *Tour through the Eastern Counties of England* as a local landmark on the road between Ilford and Romford known as 'the Whalebone').*

And whale scrumping remains alive as ever. I'd go so far as to say it's endemic to the British constitution: the enduring desire to pilfer the remains of stranded whales, both recently arrived and long-ago vanquished. *Remains alive as ever.* One of the most famous whale skeletons in England, the whale of Burton Constable Hall, is currently missing eleven pieces. The only 'real' whale discussed by Melville in *Moby-Dick*, this 58-foot sperm whale threw itself onto the Holderness coast in East Yorkshire in 1825. It subsequently found its way to the grounds of the Lord of the Manor, Sir Thomas Aston Clifford-Constable, 2nd Baronet. He in turn had the skeleton preserved, articulated and put on display in his grounds, a first-of-its-kind exhibit.† Thomas Beale

* According to Defoe, 'passing that part of the great forest which we now call Hainault Forest, came into that which is now the great road, a little on this side the Whalebone, a place on the road so called because the rib-bone of a great whale, which was taken in the River Thames the same year that Oliver Cromwell died, 1658, was fixed there for a monument of that monstrous creature, it being at first about eight-and-twenty feet long.'

† The whale skeleton fad soon caught on in Britain. When Melville visited London in 1849 (with *Moby-Dick* on his mind), and stayed at Craven Street near Trafalgar Square, he must surely have been alerted to the fact that just a few years earlier, in 1831, the 95-foot skeleton of a blue whale had gone on display outside St Martin's Church, Charing Cross (about a two-minute walk from his lodgings). According to the historian Richard Altick, the whale in question 'was displayed in a capacious tent, and buyers of two-shilling "saloon tickets" were entitled to enter the whale, even as Jonah had done, but in much

advertised the 'magnificent specimen of osseous framework' in his 1839 primer for the nascent field of cetology, *The Natural History of the Sperm Whale*. Philip Hoare renewed public interest with the publication of *Leviathan* in 2009, and the Burton Constable Foundation subsequently went about restoring a skeleton that had fallen into ruin. And been thoroughly picked over.

> Whale Bone Amnesty at Burton Constable! We are launching an appeal for the return of bones taken as souvenirs when the skeleton was left languishing in the grounds for over 100 years.
>
> @BurtonConstable Tweet, 11 June 2019

---

more comfort. A platform had been erected inside the rib cage, and here, while a twenty-four-piece orchestra played, visitors could relax at cozily placed chairs and tables', *The Shows of London* (1978), 305. The irony of this dead whale being placed in such close proximity to the Houses of Parliament was not lost on contemporary observers. In Leigh Hunt's *Tatler*, one archly satirical anonymous writer (Charles Dickens contributed extensively to this publication) wrote that 'in the open ground in front of St Martin's Church, suddenly appeared the other day, puzzling the members of Parliament, a long yellow wooden building, with its title "Pavilion of the Great Whale".' In language strikingly similar to Melville's Burton Constable whale ticket-price riff – 'Sir Clifford thinks of charging twopence for a peep at the whispering gallery in the spinal column; threepence to hear the echo in the hollow of his cerebellum; and sixpence for the unrivalled view from his forehead' – the anonymous writer suggests that it 'should not be omitted, that the price of admission to the gallery is one shilling, and to the interior of the whale, or the inside fare, two'. Having quietly established the association between Members of Parliament and whale, he ends with a parting shot, noting that 'the creature has no teeth.'

MUSEUM PLEADS FOR RETURN OF MOBY DICK WHALE'S MISSING BONES. Burton Constable Hall, near Hull in East Yorkshire, has issued an 'amnesty' for any light-fingered visitors or former staff who, over the years, may have pinched one of eleven missing parts of the skeleton of its 58ft sperm whale.

The Times, 12 June 2019

Burton Constable Hall is looking for its whale bones. The country estate in Yorkshire County, England was built in the 1560s atop a twelfth-century tower and is today a 'building of exceptional interest' on the country's National Heritage list, which hosts tours and exhibitions, re-creating Elizabethan England for visitors within its well-preserved interior and collections. The Hall achieved literary immortality in 1851, when Herman Melville described the skeleton of a sperm whale displayed on the grounds in *Moby-Dick; or, The Whale*.

Newsweek, 14 June 2019

Missing are bones from the tail, the left flipper and eleven of 44 vertebrae. While 3D printing replacement is an option, the Burton Constable Foundation hopes to recover as many of the original bones as possible.

BBC News, 11 June 2019

One of the bones is a really big one – the fourth vertebra back down from the head – a bit larger than a dinner plate, so I'm sure whoever has taken it knows where it is.

Curator Philippa Wood, quoted in the *Shropshire Star*, 11 June 2019

> I'm afraid that while we have had two sets of vertebrae brought in, neither belonged to our whale, while the two vertebrae we were informed of which had a clearer provenance (one of which has apparently been made into a coffee table) alas never made it as far as the Hall. So while I'm fairly certain that the bones are out there, I'm afraid we have not had any returns.
>
> Curator Philippa Wood, email communication, 21 October 2020

I've known them all: opportunist scavengers, private collectors, grave-robbers, thieves, Brexiteers, smugglers, suppliers, revolutionaries, deviants, hoarders, fetishists, aristocratic patrons and buyers. As with their forebears, the attraction to the corpses of whales is a complicated weave of personal psychology, committed politics, economic imperative and magic – with each thread negotiating a longitudinal warp of convention, propriety and legality.

First-timers and opportunists make up the first strand: the tourist who spots an unusual souvenir; the contract waste worker who perceives an alternative source of remuneration; the teenager who just gets a whale in their bonnet and takes a crowbar to a jaw; the person who decides to surrender to a newly-discovered and hitherto undetected micro-deviancy. There's really no particular 'type' here, and many just leave it at the first incursion. On a recent visit to an East Yorkshire beach, I saw a man walk up to a whale corpse in the early morning light, as though under the spell of one of the lesser-known *River Cottage Foraging Guides*, and confidently knock out a tooth with a hammer. 'I've never seen anything like this before,' he said when I asked if tooth scavenging was a hobby.

You then graduate to the slightly more committed individuals – my principal interlocutors in what follows: the modest private collectors like Sandy in Plymouth, Ron in Bristol or Mick in Torquay. They patrol the nation's beaches, flea-markets and auctions in anticipation. Sandy's cabinet of curiosities, assembled over the course of a lifetime, is a remarkable feat of curation, memorialisation and self-reflection, an acknowledgement of that part of herself and her life that she recognises as inextricably whale. An attempt on her part to make sense of a life defined by dislocation and rupture, a collecting of something irrevocably torn asunder: past trauma rearticulated among the bodies of whales.

Ron and Mick, on the other hand, conceive of the stranded whale in more nationalistic terms, collecting bones (old and new) as mementos of a faded British glory. This form of nostalgia, which became all-pervasive in the run-up to the Brexit vote, obviously harks back to Britain as former hegemonic naval and shipping power: Nelson, Trafalgar etc. However, it also refers to the days when Britain thought of itself as a mighty whale (one of the reasons why Thomas Hobbes decided to embody his conception of sovereign state power in the figure of the Leviathan). Extending the association, Ron and Mick conceive of the present national decline in similar terms to those of many eighteenth-century English satirists: as a beached or rotting whale corpse.

They also take things a step further, forcing us to tangle with some of the more unambiguously illegal activities of the subculture. Railing against the 'bureaucrats in Brussels' and the 'mountains of red tape' constricting 'British free trade', Ron and Mick take an oblique but discernible pride in piratically hawking their cetacean wares. For them, the act of selling the stranded British whale is tantamount to resuscitating a swashbuckling nation that once swam every ocean of the world and did battle with that great nemesis the Napoleonic behemoth or elephant.* For both of them, the UK's decision to vote leave in 2016 was basically akin to the final liberating scene of the 1993 American blockbuster *Free Willy*.

Ron and Mick are not alone when it comes to the buying and selling of whales. Neither are they at the top of the food chain. Big Blue, a patron-collector who lives in the North, supplies high-end customers for a variety of cosmetic, medicinal and ornamental purposes. According to Blue it was around the mid-eighties that the demand for stranded whale parts experienced 'something of a resurgence'. After the IWC's (International Whaling Convention's) global moratorium on whaling (passed by twenty-five votes to seven with five abstentions), 'non-antique' whale artefacts became difficult to come by. Simultaneously the world transitioned into the Save the Whales paradigm (an organisation that started up in 1977), and while this thankfully served as an ideological deterrent for most, it was an encouragement to those untroubled by financial (and therefore legal) restrictions. As raw materials became scarce and taboo, the demand increased, and so the cetaceans that threw themselves onto British shorelines arrived with a bull market price on their heads. One of Blue's more recent commissions was from a Swedish millionaire who offered him serious money

* See for example Charles Esdaile's *Napoleon's Wars: An International History, 1803–1815*: 'In May 1803, the whale went to war with the elephant.'

Engraving after Abraham van Diepenbeeck, *The Rescue of Andromeda* (1632–35), from M. de Marolles, *Tableaux du Temple des Muses* (Paris, 1655).

for a suite of whale sex toys inspired by Greek mythology. Apparently the client wanted to feel like Andromeda – or an Andromeda that Perseus failed to rescue and who was consequently ravaged by Leviathan. Only non-antique teeth would do.

From boutique black market to cetacean black magic.

There's a coven in East Anglia who, instead of 'Eye of newt, and toe of frog,/ Wool of bat, and tongue of dog', incorporate cetacean body parts into their daily rituals. Bones, Aggers and Cat (the commune's three principal members) place their cauldron on the lumbar vertebrae of a fin whale and speak of the ways in which dead whales direct their potions and prayers. They are also in possession of a jawbone, which they use as a magical gateway – or at least magic as they understand it – as facilitation, a stepping beyond the bounds of habitual existence to explore other ways of being in the world.

One thing I want to be clear about at the outset – *Strandings* is not an exposé. It's not my intention here to have anyone persecuted, fined, humiliated or arrested, which some of these people potentially would be if they came in contact with too much direct sunlight. While everything here corresponds to lived experience, I've made sure that all those precariously placed individuals who skirt and sometimes cross the margins of legality and propriety remain firmly rooted in the woodwork. In writing this I found myself negotiating something of a dilemma, unable to not tell this story, and unwilling to betray the confidences I'd built up over the years. As a consequence, a signpost here, an occasional conversation there, has been subjected to a degree of lateral drift. Sandy, Ron, Mick, Big Blue, Louise, Bones, Aggers, Clive (myself and my immediate family included) – we're all out there somewhere. In the telling, perhaps one or two have become rather 'loose fish'* – flashing up from the

* Melville's borrowed term from the whale fishery referring to his own disguises and necessary sleights of hand.

page the occasional author-reflecting surface and betraying one or two moments of creative reinterpretation – but I'm afraid it was a matter of necessity.

Also: I get it. Some, if not most, of what I've just outlined may nevertheless strike you as deeply eccentric at best – antisocial, criminal, and barbaric at worst. This is perfectly understandable – in my time I have both heard of and even stumbled upon some really shocking scenes of whale vandalism and desecration. Which I condemn unequivocally with the rest. But again, as with the whale scavengers of post-revolutionary England, I would urge a momentary setting aside of whatever indignation may have accumulated so far. Allow me to show you why the necro-cetacean subculture flickering and fraying at these islands' peripheries, made up of stranded cetaceans and their scavenging counterparts, demands our collective attention. Now more than ever. Why this weird runic correspondence between species who find themselves stranded matters.

I've lived a fairly convincing double life for some time now: my 'normal' human life and the dead whale one. I have a family, a job – perhaps inevitably I became an academic with an interest in Herman Melville (particularly the late poetry that almost no one reads). I didn't go on to collect bones or whale artefacts (honest to god, guv'nor); instead I put together *Strandings*, a book of the dead which tries to make sense of, and justify, a life lived among the stranded whales, and of those who surround and scavenge them.

What began as a teenage crush for me soon developed into something of a Goth hobby, a way to mitigate a loneliness and unsteadiness I felt; create for myself what I realise now was something of a surrogate family. I soon started visiting other whales, on beaches, in museums, archives, back gardens. In the process I gradually became aware of others like me, others like her. It turned out that many people have early formative meetings with dead cetaceans, their subsequent lives – as with so many who came before us – defined by a return to an initial interspecies moment of contact, connection and sympathy.

Harmless enough.

Then, to my increasing dismay, my relationship with the stranded whales began, slowly but discernibly, to mutate. Intensify. The shift coincided with my attempt to establish a more recognisable human family. The associations steadily spooled out of control: the gradual opening of a tomb ... whale upon whale lodging themselves, squatting, in my brain, refusing to leave. With their capacity to alter and direct the course of particular days, particular nights. Initially I assumed it was a mental breakdown, my just rewards for pursuing an admittedly morbid pastime. After a full-scale panic attack on a beach in Skegness, I began casting around for solutions, madcap schemes to try and mitigate – rationalise – their presence. I tried therapy, considered exorcism, went on a variety of purgative pilgrimages; reached out and listened to my fellow necro-cetacean enthusiasts.

Then I started listening to the dead whales; came to realise: they weren't only flexing their presence in my malfunctioning head; they were also making their presence felt 'out there'. The whales were coming ashore in their droves. The wider culture registered it too: society and necro-whales in a state of mutually triggering synchronicity and entanglement. Twenty sixteen (which at the time felt like the most momentous of years) began with the biggest stranding of sperm whales in recorded British history. The thirteen-year-old kid in me, the one who'd aided and abetted a trophy hunter without question and buried the evidence, began searching for some proper answers. Who are these people that insist on meddling with the carcasses of whales? What compels a person to saw off the jaw of a sperm whale? Why were the stranded whales suddenly asking me, all of us, to sit up and take notice?

2

Plymouth

I'd heard about Sandy from an American marine biologist called Graham. We were standing under the skeleton of a North Atlantic right whale at the New Bedford Whaling Museum in Massachusetts. I remember the moment vividly because after about a minute's chatting, I felt a drip on my head. Assuming it was a leak in the roof, I ignored it. Graham said that my hair had just been conditioned by whalebone oil. It turns out that, no matter how thoroughly you clean a carcass, the slow drip-drip can continue for decades. Every morning the museum staff wipe away one or two drops from the floor.

Graham had met Sandy only once, in the mid-nineties, but they'd corresponded on and off for years, swapping semi-regular reports of their mutual passion: beachcombing. He'd passed on my address and she'd got in touch, agreeing to show me her collection in exchange for one or two chores around the house and garden. Nothing too strenuous apparently – potted plants she sometimes struggled to lift and a rose to dig up.

Summer, 2014. A warm Saturday morning, and I was being summoned to a garage to do odd jobs for a stranger. The concrete certainty of just accepting an offer like this. Sure as the tide, I got in the car and drove to Plymouth, a port city and naval dockyard on the south coast of England that marks the border between Devon and Cornwall.

I've come to know Plymouth and its Sound fairly well over the years. It's something of a porpoise and dolphin graveyard. Marine life enters between Rame Head and Heybrook Bay (a 4 km stretch) and, from a mammal's perspective, it's much safer to keep swimming on the seaward side of the Breakwater Fort. Any further and you're almost completely surrounded: Drake's Island in front (encouraging you down the River Tamar on one side and River Plym on the other); Jennycliff Bay and Fort Picklecombe to the east and west, and the Breakwater blocking any straightforward route out.

A retreat might also be more easily accomplished if there were no traffic. Aside from the daily back and forth of the massive Santander ferries, the jet skis and motorboats, the main disorientations come from the variety of destroyers, aircraft carriers and nuclear submarines that dock at the Breakwater. Some of the cetaceans that wash up here look as though they've taken part in a one-sided aquatic sabre fight. Porpoises have been known to come ashore without a head, guillotined by one of the continually patrolling propellers often extending three times the height of a human.

Imagine the intensity of the assault on the cetacean sensorium while trapped in an echo chamber like the Plymouth Sound. The sending out of micro-pulses; the competing wall of low-frequency noise; the instantaneously received counter-signals; the deafening scream of a jet ski as it misses your body by a couple of inches; the thunder of engines enshrouding your very being. It takes the hum of a single fridge for me to go mad.

As various chroniclers of the sea have pointed out, whales (sometimes dolphins) have a very particular way of responding to sensory overload of this kind: they become 'gallied'. Melville describes this state as a 'still becharmed panic': 'that strange perplexity of inert irresolution, which, when the fishermen perceive it in the whale, they say she is gallied'. It probably comes from the Old French *galler*: 'to gall, fret, itch; also, to rub, scrape, scrub, claw, scratch where it itcheth'.* On becoming overwhelmed by a multitude of sonic itchings and scrapings, cetaceans simply switch themselves off, revert to pilot mode. I've seen a porpoise from a boat in Lyme Regis as if under a spell. A group mackerel fishing trip in the early 2000s. Our captain steered us back towards an object floating in the water we'd flown past at considerable speed. Eyes wide open, paddles gently undulating. 'Probably resting', came the announcement over the crackling tannoy before engines roared back into action.

It's rare for whales to swim into the Sound. I'm sure they can feel a sonic megalopolis such as this from hundreds of miles away and steer well clear. That said, they do occasionally get lost here. There's footage from 1965 of a pilot whale stranded as high as the St German's River. That's a left at Saltash and then a few more kilometres still – a committed inland swim. In June 2018, a family on board a motorboat filmed what they thought was a killer whale in Jennycliff Bay. It's difficult to tell from the footage: it might have been a just-as-unusual Risso's dolphin. Either way, it was one of the grampuses that have a habit of pointing a long dorsal fin proudly out of the water, as if in prayer.

Coming off the A38 from Devon's second city Exeter (my then place of work), dropping down the tangled 1960s intersection, I caught sight of the River Plym at high tide, bathed in sunlight. Three waterskiers pulled by high-horsepower engines (one of them a jet ski) were screaming down the central channel. Just a few of the headlines from recent years: 'Dolphins harassed in River Tyne prompt police

* Randle Cotgrave, *A French and English Dictionary* (1673).

warning' (two jet skiers filmed buzzing the exact spot where dolphins come up for air).* 'Dolphins "harassed" by jet skis and boats on River Tweed' (deliberately speeding towards breaching dolphins).† 'Boat owners warned after dolphins are chased and harassed' (two boats chasing dolphins in and around Fishguard Harbour).‡ I can't help thinking about what the retributive equivalent might be: maybe fitting these assailants with hearing aids, then tying them to a chair, placing that chair on a runway, and having a squadron of F-16 fighter jets perform a series of near misses and flybys. That would gall, itcheth.

I pulled into a neglected backstreet like any other: dog shit, the occasional street lamp, a surprisingly intimidating mural of a badger and a fox on a garden gate, a succession of rusting garages with moss-covered asbestos roofs. I stopped the car and waited. Seagulls cranked up their communal morning salutations, the terraced roofs they perched on white with guano. I poured myself coffee from a Thermos, a former girlfriend's really well-judged birthday present. She'd plastered it in whale stickers: right, minke, fin and sperm whales stained from years of outings.

Sandy opened the gate next to the fox and badger. In her sixties, over six feet tall with confidently applied eyeliner, she greeted me from above and thanked me for coming. Her face was magnificently sunburned and she was dressed for gardening: a soiled pair of jeans and a T-shirt celebrating the post-Grateful Dead career of Jerry Garcia. I liked her

---

* *BBC News*, 6 July 2019.

† *Berwick Advertiser*, 24 October 2018.

‡ *Wales Online*, 12 September 2016.

immediately. She ushered me in through the gate, and I followed her down a dark and thoroughly overgrown alleyway that ran alongside her garage. The garden was a patch of grass maybe ten metres long and five metres wide, guarded on all sides by high trellis-topped fencing. A rampant Russian vine, thickened by a variety of clematises and ivy, sprawled over everything and extended the height of the perimeter by several inches. A completely white cat basked at the foot of some sliding doors.

'Bruno's not mine. I'm not really a cat person. I'm just looking after him for a couple of weeks – mum's unwell. He's deaf – did you know that about white cats? Aren't you, Bruno? [Whispered.] Can't hear a thing.'

Her voice was an octave higher than I expected, though not without depth. She offered me a wrought-iron chair.

'Wait in the garden, I'll get the kettle on.'

She went inside and I inspected the garden, a little concerned not to find at least some evidence of her reputed pastime. A conspicuous lack of shells, pebbles and driftwood. Most of the strandings enthusiasts I'd come across up to that point had, at the very least, turned their patch of earth into something of an octopus's garden. Sandy seemed to have made an effort to draw the strictest of divisions between land and sea. What I was sitting on was definitely land.

When she re-emerged I politely commented on the absence; I'd expected to see more shells in a beachcomber's garden.

She'd made the decision a while back. Used to be overflowing with that kind of stuff. She found it all just accumulated, turned green, and looked sad. Each pebble an outing. After her husband died, she'd decided to mostly start again. 'You don't want these places feeling too much like graveyards, do you?'

I asked about the Garcia T-shirt. She'd followed the Dead through the Midwest for the last six months of 1972. 'It wasn't anything too special – I just experienced it mostly as addiction, heartache and sexual assault.' She'd also done it all without a trust fund. You really needed a trust fund. Otherwise the hangovers become too frightening.

'But you're still wearing the Jerry T-shirt,' I said encouragingly.

She smiled, said she'd never quite been able to let go of the voice and guitar. The whole experience was a nightmare, but it had been good to hear a thirty-minute version of 'Candyman' in the early hours of 1973. Their playing had been a miracle that night.

I started telling her about the Russian Orthodox priest I'd met in Jerusalem when I was twenty-four. In a broad Midwestern American accent he introduced himself to me as Father Lazarus. I'd asked him about his story and it turned out that he'd followed the Dead for about a decade.

Sandy cut me off: 'They're all called Father Lazarus – able to resurrect themselves no matter what. Dead Head to washed-up Russian Orthodox priest – believe me, there are thousands of them on the face of this earth. It's a surprisingly well-trodden career path. All wankers.'

We stayed with the silence for a while. I felt young; asked her what kind of work she had planned for me.

'We need to move those pots over there – I had a small stroke last year and they've got a bit heavy. And then, if you don't mind, I'd also really appreciate it if you'd dig up that rose. It doesn't like it there. Finish your tea first.'

I sipped my tea.

Sandy said Graham had given her the outline of my experience in Norfolk. 'Do you know what became of her?'

I said I didn't.

Sandy expected me to continue, asked if I'd ever tried to find out. 'I could ask my people if you like? I know a man with a comet tattoo on his arm, but I don't think he was ever a woman. It would be interesting, wouldn't it?'

'How many "people" do you have?' I asked.

'Not too many.' A robin perched on the table and looked at us expectantly before darting back into a matted grape vine.

'Shall I tell you how it all started for me?' she said, sensing my reticence.

She'd spent her summers on the Isle of Wight with an alcoholic father and a beloved sister. As the father drank himself into oblivion on the beach at Shanklin, the sisters would run around semi-feral with a gang of other children. I know that beach myself: iguanodon footprints at low tide, nuggets of amber, plastics assembled from often-incredible points of origin, jet-black trees petrified in their entirety and burnished in fool's gold. Sandy and her sister had been exploring this beach with three or four other ten-year-olds, when they came across a stranded porpoise. 'A sight that absolutely stunned me.'

A real porpoise, eyes crystal clear as any human's. One of the boys in the little gang picked up a stick and carelessly prodded it on the head to see if it was still alive. Sandy had picked up another, bigger stick and swore that if he did that again she'd kill him. 'I was overcome by an urge to help in whatever way I could. The tideline was receding – about twenty metres away from where it was lying. I asked each of them to help me carry it back to sea. Three of them helped, my sister most of all. The boy I'd threatened to kill just stood and sulked. No use. It was too heavy and, it became apparent after about two minutes, already dead.

'What did you do with it?'

'We buried it. Gave it a funeral. I said sorry to the boy; asked him to help dig. It was mostly sand and we dug the grave in no time. Maybe four foot deep; rolled the creature over and in. I recited fragments from my mother's funeral service: "Change the body of our low estate that it may be like unto his glorious body." My sister and I were burying her too.'

When you bury a cetacean on a beach, it really ought to be in early summer, when the likelihood of extreme scouring tides has subsided. Otherwise the sea almost immediately excavates your carcass and hands it you back. Sandy had buried her porpoise deep enough, which meant that it hadn't just floated up again on the next tide.

'We've lost the knowledge of how to bury things properly,' Sandy said, thoughtfully staring at her mug. A siren swirled away in the distance. 'When I came back the next summer, the situation with my father had deteriorated. He was beginning the day with alcohol now. He didn't physically abuse us, but we were starting to get neglected – unwashed, same clothes days on end. Sea urchins. The first day we arrived, with him barely able to speak by midday, we immediately ran along the cliffs to the grave. The sea had reoriented everything, so we were only able to guess at the precise spot. But then, something of a miracle. It wasn't the porpoise we'd buried, but I swear – on that section of beach – on that morning – we found the perfectly preserved vertebra of what I would later find out was a minke whale. Jammed between rocks like a lump of driftwood. When you find something like that it feels as though the beach is blessing you. It feels like a validation of character. Can you imagine how exciting that was?'

'How big was that? Substantial, right?' Finding an intact minke whale vert on the south coast – in fact, finding one anywhere, and then being so lucky as to get to it first – is incredibly rare.

'About the size of a propeller from a decent-sized boat. We could just about carry it home.'

Sandy downed the rest of her tea and looked at me. 'Right, since we're swapping war stories – your turn. After the girl, what next? How did you tumble down the rabbit hole, or whatever the whale equivalent is?'

I thought it polite to share back. It must have been 2001 or 2002. I was fifteen. Very straight-laced – in spite of the encounter; maybe partly because of the encounter. The meeting with the blue-haired comet woman had been pivotal – I knew that much. But at this stage I also felt my life didn't need any more ambiguities in it. I look back at photos of myself and I'm a Stepford son – shirt tucked into trousers, desperate to fit in, disguise any ambiguous class credentials, head down at school. I say this without self-reproach or pity – just observations. I was a fairly reactive personality, forming in conservative relation to an acrimonious divorce, and a misplaced sense of social precarity. My mum – my talented, loving, struggling German émigré mum – inherited from her parents (decent, hardworking, unlucky people): a wardrobe, some papers relating to a hairdressing shop they ran for thirty years, and some other documents regarding my great-grandmother's involvement as a low-level official in the local Nazi party during the war. My Corporate European Director father had embraced a form of extreme self-reliance in response to his own trauma (forced to become family bread-winner at eleven). She was my father's PA. That's how they met. Right 1980s by-product, me.

I remember wanting to become a lawyer, an accountant, a policeman. In spite of all the grounding, still the desperate need to tie myself to something solid. Initially I was proud of my father's working-class-no-formal-qualifications-whatsoever-successful-businessman credentials – a pride which later tapered off and turned to childish enmity as his story became structurally incompatible with my own self-image. He eventually lost all his assets and money (save his pension) after the 2008 crash. But I'll resist this constant temptation to make invisible the streams of patronage and support that have defined this particular life, such as my part-funded place at an all-boys school which greatly exaggerated my sense of chip-on-shoulder social difference. My English teacher was fond of saying that you couldn't buy class; a younger kid called Andy repeatedly referred to my parents as pikeys.

I got a weekend job at Debenhams (menswear, mostly, but they sometimes got me selling lights); used the money to travel into London. I knew this wasn't me, but I took to walking up and down Fleet Street and the Inns of Court, watching what I considered to be successful people get in and out of black cabs. I even indulged in fantasies about it being my daily commute – these people as my colleagues. For a time I wore a suit and tie (Debenhams staff got a 20 per cent discount) and experimented with buying the *Financial Times*.

Sandy looked concerned.

Then one day on the train to London I handed in my notice – fired myself in an approximation of teenage rebellion – and decided instead to head for the Natural History Museum's whale hall filled with life-sized fibreglass models (and articulated skeletons) of cetaceans swimming through the air. Exhilarating. Straight across the bridge to Embankment, St James's Park, Green Park, Hyde Park, Kensington. Half hour's walk at most. A day with the whales and then back home for dinner. All of this made possible by the then Labour government's decision to make the London museums free.

Because of the money the museum threw at its dinosaurs – creatures always lavished with the latest technology and animatronics – the whale hall itself felt distinctively old-fashioned, traditional, durable; underinvestment turning it into something of a retro sanctuary. Suited me down to the ground. Museum visitors seemed to either actively avoid it or spend a comparatively shorter time there. Soon I was priding myself on a connection to these creatures that nobody else had and that nobody else could possibly understand. I learned all the whale-related museum signposts by heart – read books, people-watched – and revised for my GCSEs. After a while I realised I wasn't the only one. A kid maybe two or three years older than me always sat in a particular corner of the hall. He invariably wore a Warhammer T-shirt partially obscured by black hoodie; made frustratingly good sketches of the various skeletons. Theo and I eyed each other, first with suspicion, before relaxing into a nodded greeting. Passing in the hall one afternoon, I eventually commented on his sketches; asked him if he knew that the skeleton in question had stranded on the east coast in the 1920s. He studied me, and then began detailing the provenance of every skeleton in the room. In great detail. I tried to keep up with my own esoterica. That provoked him to call in the cavalry. From the inside of his jacket pocket he pulled a trump card: the inner ear bone of a blue whale, given to him by his late grandfather, a 'well-known artist'.* He held it out to me; said, with an air of absolute knowledge and authority, that if I concentrated hard enough I might feel the remnant ghost-pulses of long-dead whale-song.

Fuck. You.

We became friends of sorts – I learned a lot from him. We watched and didn't understand Béla Tarr and Agnes Hranitzky's *Werckmeister Harmonies* (2000); read and didn't understand the recent translation of the 1989 book it's based on: *The Melancholy of Resistance* by László Krasznahorkai. We visited two rotting whale corpses along the Thames Estuary and Essex coast; stared at them as the bloke does at the end of the film.

And then we took a day trip, a train and the bus, down to Charmouth, the village just along the coast from Lyme Regis that both of us knew well. Theo wouldn't tell me exactly what we were going to do. We came to the intersection which leads down to the famous fossil beach and he said we weren't going to take the usual turning. Instead he led

* The only other one I know of is in the possession of the sculptor Steve Dilworth, who lives on the Isle of Harris in the Outer Hebrides.

me to another bus stop, and before long we were standing in front of the dilapidated whalebone arch at one of the garden entrances of Chideock Manor house. He explained that his grandparents had retired to the area, and that his grandfather had always taken him to this spot when he was young and told him that this was a Persephone door, the entrance to the underworld. The lore of a whalebone arch: any wish you make while passing through will be granted – just don't look back, otherwise you face eternal damnation.

I walked through, so nervous of trespassing on someone's private property that I both forgot to make a wish, and looked back at Theo. He laughed. I blushed. So be it. He brought out the ear bone again and started reminiscing about the good old days, his grandfather teaching him how to make life drawings of cetacean skeletons. As I listened, I felt myself getting increasingly upset. Explaining whales, to me, as if he had access to the eternal currents of the universe. He didn't know jack-shit about stranded whales. What they really meant – the secrets and counsel we kept. The dead whales were my inheritance – my alternative foundation. All very well receiving a souvenir from a posh grandparent; I'd been elbow deep in whale gore, mate. When he pulled his party trick, held the ear bone to his ear bone, its bass tremor swam through me. My relatives, not his. I think I politely said I needed a piss, and left him there at the arch – immediately blocked him on MSN Messenger, changed my email, stopped hanging out at the whale hall. I bumped into him again four years later, in the crowd that had gathered to watch the 2006 Thames whale. Offered my apologies for running off like that. We're still in touch from time to time. Anyway, my cetacean rabbit hole: a fusion of class-consciousness, youthful self-pity, perceived insecurity, puberty, damnation and a teenage turf war. Also, soon after the archway, I did all right in my exams – and that bought me some more time to indulge.

Sandy asked if I wanted to see the minke vert.

Without disturbing the cat, she slid open her back door and ushered me in. About a dozen houseplants designed to offset the stench of stale cigarettes and cat food. I followed her up the stairs and into a second bedroom.

A few of the bone collectors I've met keep rooms like this one. Sandy's wasn't as elaborate or extensive as some, and I didn't feel the way I sometimes have (like an anti-poaching agent suddenly happening upon a major cache of ivory), but this one still had some remarkable specimens. I saw the minke vert on a shelf surrounded by other smaller species. Propped in the corner were two ribs (the biggest about five foot in length); three porpoise skulls; a beautifully bleached shoulder blade of what must have been a fin whale.

While I surveyed the room, making appropriate noises

of appreciation, Sandy picked up a box. 'I don't know quite what to do with this.'

Sandy unfastened the lid to reveal a lump about the size of an Alsatian's head wrapped in satin silk. There's only one thing that would get stored like this (though this was surely far too big to be that). She glanced up at me, knowing what I was thinking. Then she very carefully revealed the solidified white-grey matter, encrusted with black rot.

'You are fucking kidding me.' I'd never seen a lump like this, not even in museums. Not anywhere. Ambergris: by-product of the sperm whale's digestive system turned uniquely prized ingredient of the perfume industry. 'It's not even sea-worn, is it?' So fresh the salt water seemed not to have reacted with substantial sections of its surface – as though the whale had politely coughed it up into a flipper and placed it delicately on the foreshore.

'Christ, is that a squid beak?' I pointed out a semi-digested form protruding from one of its contours.

'It's a good piece,' Sandy said, playing it cool.

The price of ambergris is always dependent on how long it's been out of a sperm whale's stomach. Too long and it becomes 'leached', the salt water dulling its aromatic range and turning it from grey-white to a solid dark, dull grey. A lump such as this, though – you could see that the surf had barely rolled it – starts to sweeten with age. Leached ambergris (the most commonly found and sold) is currently selling at about $35 a gram. This was probably a kilo's worth. You almost never get pieces as pure as this. To the right buyer, it was worth more than its weight in gold.

She'd picked it up on a Devon beach in the early noughties, 'around the time of your legal career', Sandy said. A once-in-several-lifetimes' discovery. It was just sitting there on the tideline, draped in a piece of seaweed. Every now and again you hear of someone finding a nugget. In the last ten years or so, bits and pieces have washed up, particularly on the north Devon and Cornwall coast: Ilfracombe, and Summerleaze Beach just outside Bude.* I've had my hopes dashed at a variety of locations: I see an unsightly lump or globule, my heart skips a beat, and it's always either asbestos or another form of exotically deteriorated industrial waste.

'Can we do the test?' I said, conscious of my over-enthusiasm, and knowing that she'd probably done it multiple times before. Sandy was a step ahead, and retrieved a needle and a lighter from her back pocket. I took the needle between my thumb and index finger and began heating the tip, my hand surprisingly still. I waited until the point of sterilisation and then held it for a few more seconds just to make absolutely sure. Sandy had placed the container on the shelf. I moved towards it and very carefully inserted the needle. The outer crust gave way easily – a texture of wax, honeycomb and candyfloss. I had done this once before, on a piece that was so old and sea-worn as to be basically petrified. That hadn't counted.

As hoped, a translucent bead formed at the point of

---

* See various news stories from the UK press (with varying hopeful estimates): '"Whale vomit" worth up to £40K', *Cornish and Devon Post*, 6 February 2013; 'Man finds whale vomit worth £100,000 on Morecambe beach: It's enough to make you Moby sick,' *Sun*, 4 April 2016; 'MOBY SICK: Dad and son flog "smelly rubber" beach trash for £65,000 after discovering it is mega rare whale vomit,' *Sun*, 23 August 2016.

entry. I carefully removed the needle, turned over my wrist, and smeared its warmth across my tendons and veins. I put my wrist to my nose, and breathed in the stomach bile fumes of a sperm whale. 'Literally and truly, like the smell of spring violets', Ishmael claims while processing spermaceti on the deck of the *Pequod*. I experienced this as a similarly profane freshness – a musk that in the intensity of its beauty rebounds towards and insinuates a nightmare inversion of itself. Only ambergris and the scent of my one-day-old daughter have had this effect on me.

And then, as promised, I found myself digging up Sandy's garden. After moving various potted plants to specific locations, she directed me to tackle the rose bush. She wanted it gone, which seemed a shame to me because it looked really quite impressive, the latest round of flowers still passably in bloom. I pruned it down to the base, and then began slowly prising the well-established roots away from the ground. When it finally gave way I went with it, tumbling back into the dirt.

Sandy peered into the hole I'd made, then got down onto her haunches and brushed some dirt away. Somewhat sheepishly she said, 'OK, now look at this.'

I dusted myself off, got up, and peered. What looked like one or two not insubstantial ribs were sticking out of the earth.

'Tell me that's whalebone.'

'It's my ex-husband.'

I looked at her and, for a split second, she maintained the most serious of faces.

'No, it's a juvenile long-finned pilot whale I found about

four years ago and buried. A male. The plan is to dig it up and arrange it on the table, and then maybe articulate it. Apologies for the subterfuge: I would have done it myself but, as I say, since this stroke thing I find it all slightly more difficult. I was waiting for the right moment, and you seemed like a decent candidate for the job.'

Once again assisting at the site of a cetacean grave, and this time I didn't feel particularly great about it.

'Shouldn't you have reported this?' I said, maybe more earnestly than was necessary.

'To put our consciences at ease – yes, I did. It was very badly decayed when I found it. I called the local association, and the only slight divergence in my account was that, rather than being washed back out to sea by the next tide, my late husband and I took the whale home and buried it in the back garden. But really, we already know what's happening to the planet, don't we? One more successfully reported pilot whale stranding isn't going to convince the overlords to stop plundering the earth and poisoning the oceans, is it?'

'How much digging is this going to be?'

'It was about five and a half foot long – a young one – so I'd dig from about here to here.' She indicated the patch of fairly solid-looking ground. 'It's a reasonably shallow grave – you'll have to be careful and do it with a trowel, especially around the fins – but I don't think it'll take you more than two or three hours. The soil around here is fairly lush. And how often do you get to dig up a long-finned pilot whale?'

'Not often', I conceded.

'I've bought us a couple of cold beers.'

Long-finned pilot whales are fabulous creatures, and sometimes affectionately known by cetacean enthusiasts as the Edward Scissorhands of the sea; their defining features being their elongated and multiply jointed index fingers that extend, aqua-dynamically, down to the halfway point of their bodies. This wasn't going to take us two or three hours, though. I'd be here tomorrow. And probably the next day too.

We decided to start from the tail up. This means that Edward's exhumation would begin with a series of small bones (seven to ten) tapering to a tip. We estimated where that tip might be – Sandy couldn't quite remember because the garden had changed so much. We borrowed another half-foot, and started to dig. I asked Sandy whether she felt self-conscious about overseeing the excavation of a skeleton in her own garden, particularly one that happened to have a grave roughly the size of an adult human.

'No one will bother us. We're fairly enclosed here. And also – self-consciousness prevents so many wonderful things from getting done in this world. You're digging up the complete skeleton of a long-finned pilot whale in a suburban garden in Plymouth. This is life passing before you.'

I started digging, the earth a combination of loam and earthworms. You wouldn't want to bury something like this in a clay soil, Sandy explained: it wouldn't ever break down.

After a few minutes, I paused my scraping and looked up at her – asked why she'd done this.

'I'm only going to give you an answer to a question like that if you keep digging.'

I kept digging.

Whales were the only creatures on earth who truly attempted to commit to a life both on land and sea. She admired that. The bones of Edward here were relics of a life lived initially above water. Bone structure, breathing apparatus, reproductive conventions, all first designed for land-based habitation. Then, at some stage in their evolutionary trajectory, they collectively changed their minds. They got back into the sea. At which point these fundamentally land-locked forms began readapting to oceanic conditions. So the various fractions of oil, grease and spermaceti that allowed these creatures to navigate and withstand the vastness and cruelty of the ocean, were additions – afterthoughts. Two great commitments: one to the land; one to the sea. And in those commitments they ended up slightly ill-suited to both environments, never wholly sure as to which arena they actually belonged. Sandy swore she'd seen a whale rescued by the tide that then obstinately attempted to remain on the sand. Four or five times the flood tide carried this creature back out to sea, and each time it insisted on coming back onto the beach, as though it were sure that here is where it belonged. A stranding is a momentary urge to retract something that has become absolutely non-retractable – an instance when a whale discovers that both the home and it have been irretrievably altered. Sandy knew that a similar dynamic had defined every stage of her life – for all sorts of reasons: never quite belonging. Also, if you think about it, strandings are most likely to have been the key evolutionary event. What do you think legs initially evolved to do? Help aquatic creatures escape back into the ocean when they ran aground. We have legs because of that basic emergency. Sandy pointed at her pins. 'Anti-stranding devices – that's all these are.'

She paused, and took a deep reflective breath, the inhalation of a smoker ruefully experimenting with the desolation of a long-term future breathing mere air. 'Then, like you, there's a whole load of displacement and surrogate family stuff going on too.' Officiating over the burial of the porpoise as a girl – a way of working through her mother's death. Mainly, though – and she apologised if this was quite predictable – the stranded whale was probably her father. A daddy complex. Washed up, beautiful, full of promise and life, and utterly helpless – not quite ever knowing who he was or where he belonged. Floundering on the sand. 'I'm afraid that's the heart of it. I'm just another grieving Isis, collecting the scattered and mutilated bones of my beloved Osiris – trying to stick him back together again.' Sandy looked over into the hole I was digging and whispered, 'Hello, Daddy.'

I smiled and carried on digging.

An hour in, and I finally located the delicately designed tip of Edward's tail. Some of the cartilage sticking the bones together had been preserved. At this early stage, I thought perhaps that I'd be able to lift sections of the spine and fins instead of individual vertebra and paddle bones. No such luck: the next fifty came out one by one, the saving grace being the solidity of the bone itself. Buried underground like this, it might start breaking up, you'd think, but I was able to lift each one out of the dirt quite confidently without fear of causing damage. I began each bone by excavating the protruding spinous process of a particular vertebra (the point of attachments of ligaments and musculature), and then simply tugged. Most came away relatively easily, apart from the odd piece that was serving as the anchor for a rose root. After removing each bone, I handed it to Sandy, who removed the dirt with the brush of a dustpan before conveying them inside. I'd decided not to look at the arrangement until I'd finished my day's work.

Two hours in – the heat of the day – and I started searching for the fins.

Some time after helping Sandy in her garden, I came across the work of an artist based in Iceland called Marina Rees. One of her pieces is entitled *Bones of a Long-Finned Pilot Whale* (2016), a collection of pamphlets and art installations that detail the process of recovering a stranded whale and then preserving its skeleton. A short film, directed by Justin Batchelor and entitled *Marina and the Whale*, shows her undertaking all those processes that are either ignored or erased by our otherwise entirely flesh-dependent society: butchering or 'cutting into' the blubber; removing organs; boiling and treating the bones; scraping away any remaining pieces of sinew with a knife. Marina then stands her perfectly picked ribs and vertebrae against the walls, and carefully lifts the skull to camera. Rees is brilliant at showcasing the familiarity of something that initially registers as alien or strange: here is one of her social media posts from 2016, captioned 'Working out the puzzle of a long-finned pilot whale fin bones'.

A puzzle, and certainly the most fiddly part of the excavation, but to see these bones up close – to handle them and arrange them – you are left with the shocking strangeness of how intimately related we are as species. The delicacy of those two opposing fingers: the digits that would, in humans, develop into opposable thumb and finger; in whales, digits that were now largely redundant, shaped to slice and glide more precisely through water. A single form cascading into divergent futures.

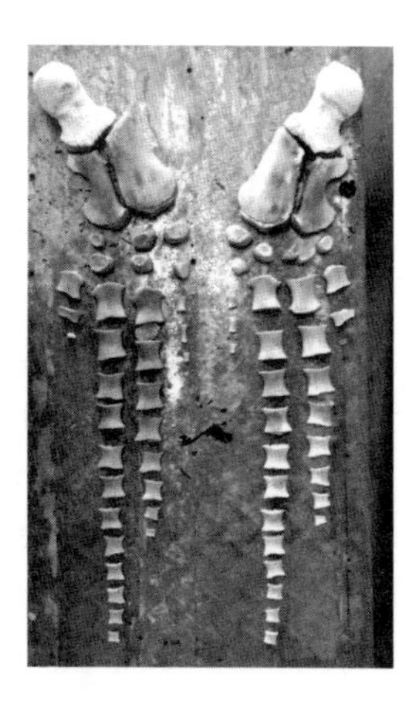

Sandy said there should be between thirty-four and thirty-six bones in each fin depending on carpal fusions and anomalies. I removed each digit meticulously, and, I think, both fins in their totality. There were so very many fiddly little parts to these fins that, in light of Sandy's soliloquy, I came to think of them as the hands of a decadent dandy, entirely unsuited to the labour currently being undertaken by me on its behalf.

We were both impatient to see the skull. I made a trench around its circumference and carefully worked my way towards the bone. I then started carefully prising and lifting. It came away, the soil caking a slightly opened jaw. Sandy handed me the brush and I revealed a nearly complete set of teeth. I then transported it through the patio doors and into the dining room. Sandy had laid out the spine and arranged the fins. Spaced out like this, about 7 ft long, Edward was beginning to look really quite impressive. I knew where the skull went.

Sandy relayed her intended preservation process. She'd wash each bone with a light washing-up solution and then begin bleaching. This destroys any impurities that might, over time, start compromising the structure of the bone. She showed me three pamphlets she was using as a rough guide. I hadn't seen these before: Lee Post's – aka 'the Boneman's' – *Bone Building Books*: Volume 1, *Articulations of a Porpoise Skeleton*, 2003; Volume 2, *The Sperm Whale Engineering Manual*, 2004; and Volume 3, *The Whale Building Book: A Step-by-Step Guide to Preparing and Assembling Medium-sized Whale Skeletons*, 2005.* Post lives and works in Alaska, offers courses on skeleton articulation, and has mounted several public installations, including the immaculately put-together fifty-five-foot sperm whale at El Refugio de Potosi in Mexico. Sandy and I have promised ourselves a pilgrimage one day. Not only is his work on a different plane, but he is one of the very few members of the international subculture to have emerged from the shadows into the realm of legality and even respectability. His most recent project was a remarkable articulation of a dwarf sperm whale in the Galapagos Islands in 2019, commissioned by the California Academy of Sciences. In spite of her admiration, Sandy expressed a slight ambivalence towards his 'Boneman' nickname, which was a bit 'metal' for her taste.

It was now late afternoon, and I was starting to feel it in my lower back. We decided to leave the shoulder blades and ribs for the following morning. Sandy waved me off. 'Promise you'll come back. I don't have anyone else to help

---

* See also his work on pinnipeds, birds, moose, bears, dogs and small mammals: https://www.theboneman.com/bone-building-books

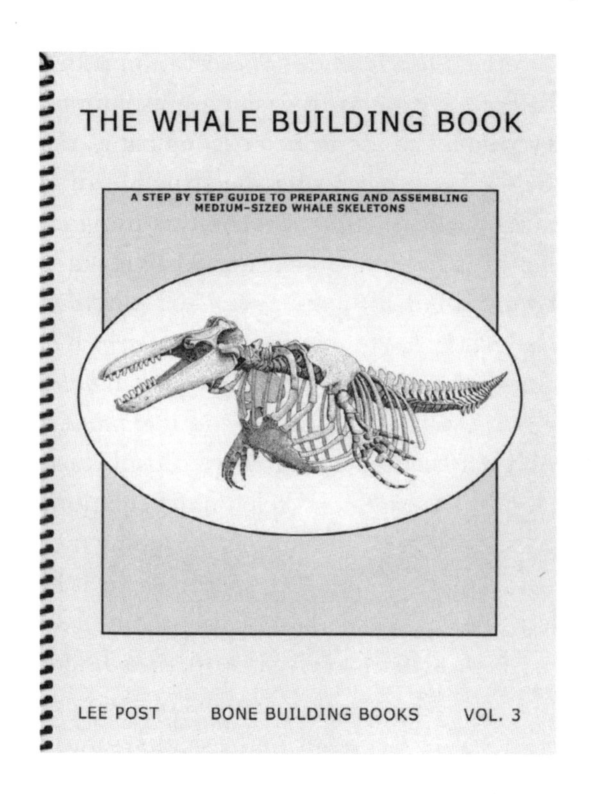

with this project.' I promised I would, and drove the fifty minutes back home to Exeter.

That evening, I gave the outline of my day to my partner Alicia. I'd told her a number of relatively eccentric stories before – my semi-regular trips to meet and greet whales on arrival (she'd joined me for some) – but never related one quite like this. As usual, she took it in her stride. Noticing the book I was flicking through (I was double-checking for long-finned pilot whale skeletons), she said that it was good to hear there were a few 'lady Redmans out there too'.

Redman. Nicholas Redman. He's up there with Post.

One of the great unsung figures of British whaling and maritime culture (never mind the necro-cetacean subculture). His *Whales' Bones of the British Isles* (2004) is a breathtaking labour of love – more Melville than Melville. Anyone interested in the bones, particularly the grand whalebone arches that are scattered across these islands and beyond, needs to get hold of his meticulous (self-published) book-catalogues.* Many decades in the making, they are beautiful works of research and curation. A trip to any part of the British Isles (or many parts of the world for that matter) is immediately enriched by taking a copy along. I have been on

*I can't help but smile at the wilful awkwardness of his various *Whales' Bones* titles (love that apostrophe): See Vol. S1, *Whales' Bones of the British Isles, Supplement 2004–10* (2010); Vol. 2, *Whales' Bones of Germany, Austria, Czech Republic & Switzerland*, (2010); Vol. 3, *Whales' Bones of The Netherlands and Belgium* (2010); Vol. 4 *Whales' Bones of the Nordic Countries, Central and Eastern Europe*, (2013); Vol. 5 *Whales' Bones of France, Southern Europe, Middle East and North Africa* (2014); Vol. 6 , *Whales' Bones of Australia, New Zealand and the Pacific Islands* (2016); Vol. 7, *Whales' Bones of the Americas, South Atlantic and Antarctica* (2017); Vol. 8, *Whales' Bones of Africa and Asia* (2019). See also the very enjoyable *The Travels of the Oostende Whale Skeleton 1828–1901* (2015). All self-published by Redman Publishing. He's the very embodiment of a particular character in *Moby-Dick*: a 'Sub Sub'. In the odd prefatory materials right at the beginning of the novel (even before the narrator bids you call him Ishmael), author-Melville affectionately introduces a character and fictional counterpart he describes as a 'sub-sub librarian'. He claims this Sub Sub did most of the grunt-work collecting of raw material (whale facts) for the novel: a 'mere painstaking burrower and grub-worm', he's 'gone through the long Vaticans and street-stalls of the earth, picking up whatever random allusions to whales he could anyways find in any book whatsoever, sacred or profane'. That's Redman.

more 'Redman retreats' than I can count; following in his footsteps, tracking down bones in the strangest of locations: forests, churchyards, hilltops, public and private gardens. The bones that litter the world. Maybe *Whales' Bones* is not quite as exhaustive as it sometimes claims to be (the bones of these islands will always be too numerous to record – as is tacitly acknowledged in the supplement he brought out in 2010); and maybe it's sometimes a little too preoccupied with monumentality (it doesn't waste its time on teeth, or the smaller, more delicate species – the Edwards of the sea). But nevertheless, *Whales' Bones* is an invaluable resource for any strandings enthusiast. Many of the adventures included here are inevitably routed through this work – and 'You won't find this in your Redman' has become something of a subcultural in-joke and affectionate refrain.

Alicia keeps asking me why (having read basically everything he's ever written, even his non-whale-related work on the brewing industry and the transporter bridge) I insist on maintaining a semi-stalker distance instead of cultivating a more collegial relationship. It's not that he's another Theo – I've attended a few of his talks, and even sent him a couple of messages (first an email, then a letter). No response. Maybe I came across as a bit keen and a bit weird (maybe it's also sometimes better to leave your heroes alone). To save a bit of face I think I transformed it all into a quiet generational rivalry – taking his reluctance as a sign I needed to forge my own path. Ah, that's not quite right either. There's another reason. Mainly it's the book's dedication: 'For Ed', the obviously beloved son who accompanied him on his research trips. Ed appears looking happy in so many of the volumes' bone photographs – he was brought along to convey a sense of scale, standing near an arch here, a jawbone there. So not only are these book-catalogues of immense scholarly import; they also narrate an idyllic love story between a father and son.

The following morning I set off at 7 o'clock, wanting to beat the seasonal Cornwall traffic. A clear ride down. On arrival, Sandy gave me a delightfully strong and sweet Turkish coffee and some baklava. I set to work on the twenty-four ribs. They, too, came away easily.

I lifted the last piece of Edward, handed it to Sandy, and stuck my trowel in the earth. I'd exhumed a whale.

Or, as it turned out, I had almost exhumed a whale. Both Sandy and I stared at the small mound of earth that had been protected by the rib cage. 'Shall we?' Sandy was more sombre than I'd seen her up to that point. I suppose it'd always been quietly understood between us that the excavation was really only building to this moment. In a further twenty or so minutes: two lighters (one with the naptha still flowing inside it), three bottle caps (one Sprite, one Coke, one anonymous white), part of a foam cup, a ball of fishing line, a Mars bar wrapper, a very faded Snickers (?) wrapper, a bright yellow sand eel lure with barbed hook buried inside it, a three-pinned piece of Lego, a mini ice cream spoon, part of a yellow plastic bag, part of a black plastic bag, three cigarette butts, part of a green plastic bag, a fluorescent orange squid lure with vicious-looking spikes encircling the base, two short lengths of wire insulation (one blue, one brown), and sixteen miscellaneous fragments: four green, three black, one red, one orange and seven white.

About half a kilo's worth – maybe not quite enough to be the cause of this particular stranding, but having all this in his stomach would have certainly slowed Edward up.* Sandy and I remained mostly silent during this final stage of the excavation. She collected every piece of plastic, as though they were the bones themselves. I followed her into the house a final time. She carefully removed some ribs – spaced the pieces of plastic neatly apart – and then resealed the vault.

Sandy went to the fridge and took out another couple of beers.

'To the end of the world', I said.

'To family', she said.

She reiterated her offer to ask around about the blue-haired comet woman – said that for a person who cared about alternative and extended families, maybe I ought to think about tracking her down; that it might do me good to do some digging myself. Surely I was curious to find out what became of her? Why she'd done it? She knew people in the area who were well connected – knew people in that part of the country.

I paused for a moment or two, scanned my feelings about it. Truth be told I'd always felt decidedly ambivalent about the possibility of seeing her again. I was curious – had even made some tentative enquiries myself – but to no avail. Took that as something of a sign. Really, what purpose would it

* Are there any plastic-free whales left in the ocean? On a Scottish beach, a necropsy of a sperm whale found 100 kg of the stuff; on a Sardinian beach, a pregnant whale washed ashore with 22 kg in its stomach. The endless and endlessly upsetting list goes on. And that was only in 2019. See 'Pregnant whale washed up in Italian tourist spot had 22 kilograms of plastic in its stomach,' CNN, 1 April 2019, and 'Whale dies with 100 kg ball of plastic trash in its stomach,' CBC News, 2 December 2019.

serve, other than risk disillusionment and disappointment? As an unknown quantity, Blue Hair could continue to fulfil the ambiguously inspirational role she'd played in my life, joining the dead whales as part of my preferred nativity story – conveniently positioned as another long-lost relative and preceptor rather than as an unwieldy autonomous 'living' presence (in this regard she joined Redman). Family members are easier to deal with in their absence. I was perfectly happy and willing to resurrect someone else's cetacean relative, but it didn't necessarily follow that I should go excavating my own foundations. Let sleeping whales lie.

3

Fishguard

A few months later, gathered outside a community beach hut at the end of Park Point in Duluth, Minnesota, the longest freshwater sandbar in the world. Lake Superior perfectly still and aquamarine blue on a summer's day. I am surrounded by Alicia's American family, my mother and a small group of friends. We'd chosen the readings and order of service in a pub on the edge of Dartmoor. I didn't think anything of it at the time, but Alicia had suggested we include a whale or two – maybe a reading from *Moby-Dick* or Thomas Beale's *The Natural History of the Sperm Whale*. I remember my anxious tone: this was going to be a freshwater ceremony; it'd be alienating to start talking about dead whales. Maybe better to leave them out this time? I was keen to make a good impression. In spite of the overwhelming hope and joy of it all, maybe just the faintest background apprehension that this was happening on a beach and that someone born in this town once sang, 'We sit here stranded, though we're all doing our best to deny it.'

We'd decided that A's father, a wonderful and gentle man with plenty of patient calm and good humour, should preside over the wedding ceremony – supported by a cast of American aunts and uncles, nieces and nephews giving readings and musical recitals. The Shaker tune 'Simple Gifts' on the penny whistle and fiddle. Everything went swimmingly. And then our priest accidentally married his daughter to a man called Peter Jackson Lothar *Ripley*.

Everyone laughed politely. All good. My full name is a bit of a mouthful. Jackson after the American singer and Eagles abettor Jackson Browne (my mother's choice); Lothar after my hairdressing grandfather.

Ripley for Riley, though.

A whale had drifted up and was gently nuzzling my brain: 'Ripley's whale'. An almost physical presence swimming behind my eyes. Ripley's was one of five sperm whales that beached in the Covehead area of Prince Edward Island, Canada (just north of Nova Scotia) in 1988. They'd met a specific and tragic end. A local rescue team came up with a plan to drag them all back out to sea by their tails. Nooses tied, the dragging began – dislodging their spinal cords and killing them instantly. They washed back up on the next tide, victims of the first cetacean mass hanging. Ripley's Believe it or Not! museums, an American entertainment franchise (with an outlet in Blackpool, UK), purchased one of the whales (37-foot and 20-tonnes) and buried it in the hope of cleaning the skeleton for future display. The earth they chose was heavy clay – not Sandy's loam – and it refused to decompose. So it lay there in state: the immaculate uncorrupted saintly carcass of Ripley's Whale.* A new sensation, a stranded whale come to visit me. Splashing around in the head of Peter Jackson Lothar Ripley on his wedding day – as he exchanged his vows.

And Ripley was only the first in a gathering wave of beached, embalmed and mutilated whales that began troubling the psychological and physical borders of my life. A slow turning up of the volume, as though they'd been provoked, activated.

Maybe it shouldn't have come as a surprise that, just as I

* It was actually discovered again sticking out of the clay in 2019 by Johanna Kelly, who was walking the shoreline and planning a beach clean-up with the Kensington North Watersheds Association.

was attempting to piece together a human family, my necro-cetacean relatives should begin acting up in this way. I'd spent a good portion of my life meeting, greeting – researching them in detail; now they were quietly coming home to roost. It happens with most forms of dependency. The gradual handing over of agency – a tipping point whereby the illusion of self-control gradually segues into the realities of subjection. The joy of marrying the person I love always just slightly offset by these repeated mental incursions. Whale corpses coming and going as they pleased, spreading themselves out as if they owned the place. *Ohrwurm*: earworm – the German word for the song you can't get out of your head. Standing at the kitchen sink, preparing a meal, having a conversation – doing anything, really – all while telling the burrowing dead-earwhales, quietly but firmly, 'Please stop.'

I was still functional: I could be a partner, do my work – but it felt like an unfastening of a carefully curated personal equilibrium. No longer a diversion, but the diversion. On any given day, Ripley's might be joined by any one of the many whales that had populated and defined my past. I was in my late teens when I'd first heard about – did some reading up on – the various embalmed whales that once toured the British Isles, transported on specially designed lorries and fitted with their own refrigerators. Triumphalist displays, really, so confident in the progressive nature of human innovation and industry that the formalin-riddled corpses were exhibited alongside the explosive harpoons that dispatched them. Perhaps triggered by the Ripley's Believe it or Not! connection, 'Jonah', 'Goliath' and especially 'Eric' began making regular visits.

Eric was the first. Toured the North in the 1930s. Named

after an almost instantly obscure pop-culture reference, Frederic W. Farrar's novel *Eric, or, Little by Little* – a morality tale about the slippery-slope ruination of an English schoolboy who dies a repentant death after running away to sea. My mind spitting up these semi-digested, long-forgotten materials.*

> 50 TONS OF PICKLED WHALE. An Exhibit for London Christmas Circus. Fifty tons of pickled whale, contained in the world's biggest packing case, were unloaded from the American steamer *Lehigh* in the Royal Albert Docks yesterday. The whale is called 'Eric' because he has to be moved little by little.
>
> *Belfast News-Letter*, 1 December 1931

> THE WHALE'S PATH. It will be midnight before the lorries which carry him reach the broad thoroughfares through Whitechapel to the City and thence past the Bank to Shepherd's Bush, which is not far from Olympia ... He was harpooned in the Pacific two years ago by a harpoon fired in safety from a steamer – not getting even a whale's

* I'd also come into contact with them again via the work of Philip Hoare and Steve Deput. Maybe it was also partly triggered by the Captain Boomer Art Collective, who had just dreamed up the twenty-first-century equivalent of the whale on tour. In 2013, they transported their hyperreal model of a beached sperm whale to London's Docklands for the Great Greenwich Whale Project. Dead whales were in the air. Have always really appreciated the Collective's justification: '[A whale stranding] stirs and mobilises a local community. During our beachings, we see an intensive interaction among the crowd. People address each other, speculate and wonder. They offer help and ask for information.'

chance. He was then about two hundred years old, and in his youth he must have hobnobbed with whales which had caught Capt. Cook's eye through his spyglass.

Yorkshire Post and Leeds Intelligencer, 11 December 1931

ERIC'S TRIP TO THE NORTH. Whale on the Road for a Week. Special police arrangements have been made along the 300-miles route from Southend-on-Sea to Morecambe for transport of Eric, the 65-ton whale that has been on exhibition at Southend for two years. Eric leaves Southend to-day on a special lorry of 180 horsepower and trailer 100 ft long. Eric's average speed will be four miles an hour.

Sunderland Daily Echo and Shipping Gazette, 26 February 1934

WHALE IN MORECAMBE. 'Eric' the Whale has arrived at Morecambe after many days wandering on the roads between Southend-on-Sea and Morecambe. Could 'Eric' speak, we feel sure he would tell us of his delight at finding a resting-place for a few months on a Fair Ground, and also express words of appreciation at the civic reception afforded by the Mayor (Ald. J. C. Wilson). It is sincerely hoped that 'Eric' will attract many thousands of visitors to Morecambe, and be some compensation for the lack of suitable swimming baths.

Morecambe Guardian, 16 March 1934

MORECAMBE'S WHALE FOR SALE. Who wants a whale? Not an ordinary sort of whale, but the only one, so far as history reveals, to have been accorded a civic reception! Once a fortnight his skin is polished with paraffin wax to keep it soft. Polishing Eric is a different proposition to washing the baby.

Lancashire Evening Post, Saturday 12 February 1938

A SAD TALE. When Eric, the most famous whale in the country, was brought to Morecambe just five years ago, he was given a civic welcome. His progress by road from Southend to Morecambe was recorded almost daily in the National Press. Now Eric has gone. He met an inglorious end to a glorious career when he was dismembered and burnt last week. The *Guardian* is informed that most of the body was destroyed at the Corporation destructor.

Morecambe Guardian, Saturday 4 March 1939

Table No. 5.—Whaling Results for the various countries in 1930/31 and summer 1931.

Countries.	Species of whales caught.							Oil production.	Expeditions.		
	Blue.	Fin.	Hump-back.	Sei.	Sperm.	Others.	Total of whales.		Shore sta-tions.	Float-ing fac-tories.	Catch-ers.
								Barrel = 1/6 ton.			
Norway	19,262	6,157	405	95	32	[1]) 1	25,952	2,316,962	5	29	160
British Empire	8,452	4,054	240	117	156	-	13,019	1,131,231	3	11	78
Argentina	599	519	30	22	3	[1]) 1	1,174	.88,154	1	1	9
Denmark	906	129	11	-	-	-	1,046	84,995	-	1	6
United States	367	165	4	-	-	-	536	49,360	-	1	3
Japan	20	337	70	418	283	[2]) 19	1,147	16,274	-	-	20
Total	29,606	11,361	760	652	474	21	42,874	3,686,976	9	43	276

[1]) Right-whale. [2]) 11 grey-whales and 8 right-whales.

International Whaling Statistics III, edited by the Committee for Whaling Statistics appointed by the Norwegian Government (1932).

Whales on tour mark the height of whaling in Britain. Eric's appearance heralded the revival and acceleration of the industry. Here are the figures from the year he made his debut and the harvest tripled.

It's no coincidence that touring whales came from Norway, the whaling industry leader from the 1920s to 1960s. The British-Dutch corporation Unilever was its primary (sometimes sole) buyer, blending de-fishified whale oil into its soaps and margarines. Unilever was founded in 1931 as an international fats conglomerate. Whale oil was the cheapest edible fat, and by the end of the decade the bulk of commercial whale oil was devoted to margarine production.

> BRITAIN TAKES THIRD OF ANTARCTIC WHALES. Three British expeditions will arrive back with whale meat for the table, whale oil for fats, whale-liver oil for vitamin A, and sperm oil for lipsticks, face creams, and candles ... Altogether, British whalers employ about

The UK's first branded margarine. 'Builds glowing health naturally.'

> 4,000 men, and the foods and fats which they contribute every year to the national store-cupboards are worth more than £10,000,000.
>
> *Daily Mail*, Wednesday, 13 April 1949

Then it all apparently started ramping up 'out there' too. New Year, 2016 coincided with the largest mass stranding of sperm whales in recorded history. I didn't know whether to take comfort from the apparent synchronicity or not. The year of the Brexit vote, the year of the shock American presidential election. It began with thirty sperm whales washing up at various locations along the southern North Sea coast. A cetacean apocalypse. Five on the German coast between 8 and 12 January; five on Texel Island in the Netherlands on 12 January (these were alive when they came ashore, though didn't see out the night); another in Germany on 13 January;

another along the Dutch coast on 14 January; one at Hunstanton on 22 January; three at Skegness, 23–24 January; one at Wainfleet, Lincolnshire, 25 January; eight at Wadden Sea National Park in Germany, 31 January (one alive); one on the French coast, 2 February; two more in Germany, 3–4 February; another at Hunstanton on 4 February; and finally one on the Danish coast on 25 February.*

When the Skegness news broke, I was in a meeting, half-listening to an enthusiastic educational consultant called Luke do some 'blue-sky thinking' on the future of 'flexi-learning'. He waved his impressive arms confidently as he spoke. The main takeaway: if you add an F to the beginning of Luke's name, it becomes Fluke. Luke waving his

* See L. L. IJsseldijk, A. van Neer, R. Deaville, L. Begeman, M. van de Bildt, J. M. A. van den Brand et al., 'Beached bachelors: An extensive study on the largest recorded sperm whale *Physeter macrocephalus* mortality event in the North Sea', *PLoS ONE* 13 (8), 2018: <e0201221. https://doi.org/10.1371/journal.pone.0201221>

flukes. I looked down and my phone read: 'Multiple Whales Stranded on North-East Coast.' I'd been following the news coming out of Germany and the Netherlands – I'd heard about Hunstanton the day before. What was happening?

I parked up and made my way towards the alien forms slumped across the beach. The whales had stranded so close together, an act of solidarity and defiance, victims of a whale Pompeii clinging on to each other in the final seconds of consciousness, affirming their fundamental sociality to the last. They were facing the same direction when they died, the intimate sonic contact of a final whispered goodbye.

Walking towards them, I had this curious feeling of having somehow been summoned – as though these corpses were exerting a gentle gravitational pull on my body. The whales attempting to live an (after-) life through me. With 'me' as the host, or whatever parasitic form 'me' was slowly mutating into. All the dead cetaceans I'd hitherto visited: had I been possessed? Was my corpus now the site of some cetacean occupation? Was I here at their bidding, potentially to pay homage to fallen comrades? Maybe I needed to be exorcised. Nip this in the bud. I knew a priest – attended his ordination at Southwark Cathedral on the south bank of the Thames. Jim'd evict them.

The now-infamous graffiti had already been sprayed when I arrived: 'CND' (Campaign for Nuclear Disarmament) and 'Man's Fault' on the tails of whales. 'Fukushima RIP' on an upturned stomach.* People were barracking each other for taking selfies. A bloke gesticulated at one of the whales and

* This is about the time and place where Daryl Hind, the whale vigilante of Chapter 1, caught the nautical dentist in the act.

shouted, 'Piss off back to your own country, mate.'* Security guards stalked the corpses and justified their gig with arbitrary and petty displays of power, 'Get back' the consistent, aggressive refrain. Assembled journalists and camera crews elicited stock moral outrage from interviewees. I stood in front of the 'Man's Fault' whale, trying and failing to shut out the noise.

Then the panic attack replete with visions: transported back to the beach in Duluth again – the blue of Lake Superior lurid green – a mass of brass and strings indistinguishable from white noise crashing in like a storm front. Tides of pins and needles marooning my limbs. The saturated outlined auras of beings squirming on the shoreline. On the horizon, a model whale – of Moby-Dick, immaculate Ripley, pickled Eric – dragged through boiling ocean by a noose. A

---

* Britain has a long tradition of projecting xenophobic anxieties onto beached whales. See for example the 1645 pamphlet *The Sea Wonder* by Anon. In the summer of 1645, a group of sailors on board the merchant ship *Bonaventure* spotted a whale in the English Channel being harried by a gang of other fish. 'In a most violent manner they did beat against [it], makeing a most hideous and a fearful noyse and falling upon her, she made what haste she could to get from them.' Taking their cue from the fishy lynch mob, and conflating the Jonah myth with anxieties concerning imminent Catholic invasion, the sailors enthusiastically joined in. According to sources, the whale was eventually driven ashore three miles from Weymouth – 'where being opened there was found in the belly of it a Romish Priest, with Pardons for divers Papists in England and in Ireland'. See also the later cartoon 'Intended bonne farte raising a southerly wind' (20 February 1798). General Louis-Alexandre Berthier feeds Napoleon in Calais, who spews an invasion army, replete with guillotines, across the channel out of his behind. An extension of Napoleon's anus, a whale beached at Dover disgorges the mathematician Gaspard Monge and invasion troops. Assembled Whig leaders and conspirators greet them cordially: 'How fragrant is this southern breeze.'

silhouetted figure, not human, riding the whale's back and pneumatically stabbing, a spider murdering its prey. The dead whale in front of me a glowing superimposition – an atomic radiation against a slate-grey English firmament. I lay down on the sand.

'You all right, mate?'

'Yes, fine, thanks.'

I lay there for an hour – gallied.

I've always tried to be as open as possible with Alicia about the extent of my interest in whales. I'm a Herman Melville scholar, after all. Comes with the territory. She'd accepted the fascination early on, even appreciated it as a fellow devotee to collecting and archives. And to be fair, such an interest, eccentric though it is, doesn't preclude me from being a good person or a good partner. We all have our quirks – this happens to be mine. For better for worse, for richer, for poorer, in sickness and in health. I told her about the stranded whales the first time we ever met – in Boston. We were both giving papers at the American Literature Association conference: hers was on pre-1917 American socialist novels (she's read every out-of-print one of them; at that time, she was working as a community organiser in Pittsburgh). Mine was on 'A Bower in the Arsacides', a chapter in *Moby-Dick* when Ishmael describes his visit to the skeleton of a stranded sperm whale claimed and displayed in a palm grove by one King Tranquo:

> Now, amid the green, life-restless loom of that Arsacidean wood, the great, white, worshipped skeleton lay lounging—a gigantic idler! Yet, as the

> ever-woven verdant warp and woof intermixed and hummed around him, the mighty idler seemed the cunning weaver; himself all woven over with the vines; every month assuming greener, fresher verdure; but himself a skeleton. Life folded Death; Death trellised Life; the grim god wived with youthful Life, and begat him curly-headed glories.

Ishmael, then in the employ of an Algerian trading ship, goes on to calculate the whale's dimensions and has them tattooed on his arm, 'as in my wild wanderings at that period, there was no other secure way of preserving such valuable statistics'. Alicia didn't bat an eyelid, not even when over breakfast I impulsively pulled out one of my myriad little black notebooks filled with dead whale notes.

The panic attack had shaken me up, though. The situation was escalating and needed addressing – before things got out of hand. Alicia hadn't signed up for this. Eccentric hobby, intellectual diversion; not necro-cetacean breakdown of newly wed husband. I'd call Clive, Sandy – maybe even Theo. Were they experiencing something analogous? This weird intensification? On second thoughts, maybe I should leave these people be for a while.

The Skegness episode prompted me to sign up for a brief course of analysis – three sessions. I went in knowing I couldn't afford to make it a regular thing. Nearly a hundred pounds a week. The analyst listened to me. About my unruly relations; about the beach in Norfolk. In a state of mounting frustration I asked for a key term, something I could just take away with me. I'd like to be fixed, please; I'd like the whales to be put back where they belong. She patiently explained that any course of analysis would always take time. I might experience moments of progress, break-through, but that there would be no 'key' or 'fix'.

'Maybe this is a kind of cetacean cathexis – gone wrong?'

She waited patiently for me to get this out.

Freud's *Besetzung* – an occupation or interest. A person, object or thing upon which you fixate: an object of desire. I might have cathexis – gone wrong – whereby the particular object starts taking over.

She said that cathexis might be something we consider, but again, this wasn't really about a specific diagnosis. It would be a working things through, a process. She told me, with more frankness than I expected, that part of this process might involve letting go of my current attitude to the conversation.

I looked at the floor; asked whether, when I told her about helping to load a sperm whale jaw into the back of a Volvo, she thought about the classic Freudian terms: castration complex, fetishism, misogynist Melville scholar fantasy sex-scene – jaw/dick into Volvo/*vulva*.

The Volvo had occurred to her and, to answer my question – no, she said, analysts didn't sit around thinking about how a patient's narrative fits into a particular Freudian mould.

'You must think it's a fairly conventional Oedipal fantasy, surely? Son helps surrogate androgynous mother figure castrate the father lying on beach before ultimately becoming unavailable to him. Something like that?'

Once you start thinking in terms of the classic patterns, she replied, it's quite difficult to un-think them. She asked me if there was perhaps another place I'd like to start. A place I hadn't saturated quite so intensely with DIY Freudian readings.

We sat there in silence again. For minutes.

There had been other whales. A whale in particular. Before comet woman; one that had stranded on the Pembrokeshire coast in 1954. I first saw it when I was five years old, sitting on the living room floor of my childhood home. All I have is a series of visual and audio stills. They flash white water and anger, pale strawberry jam and melted butter, my parents screaming at each other in the kitchen, accompanied by high-drama strings and brass.

In 1990, my father brought round a LaserDisc machine, a technology that had already long-been eclipsed by VHD and VHS. The standard LaserDisc was twelve inches across and made up of two single-sided aluminium discs sandwiched together in hard plastic. An eternally pristine fossil. To me, as a child, it represented the future. LaserDisc. He'd always arrive with some corporate gizmo – a digital clock, a quartz radio, a giant polystyrene die. Always something – destined for landfill. This was his second divorce, and he was remaining steadfastly supportive of this mother and child; his loyalty and love often manifesting in these weekly gifts, and (to his credit) regular as clockwork payments.

We had a loud and heavy wooden gate at the side of our house that must have been about fifty years old. Twice the height of me. Each time the wind picked up, or if we opened it, it came off its hinges. Opening it involved scraping and heaving it along our gravel path – making an incredible din – and then propping it up against the wall. Really the last thing you needed to deal with when you had shopping, a child in tow – on top of a full day's work. Each Saturday the same ritual, just before we set off to 'Nana' in Watford: my father would 'fix' it. He'd ask for nails and a hammer, and my mother would dutifully give them to him, a last vestige of cooperation between them. Three or four were usually enough to hold it in place for a further two or three days. The banging was loud and disturbing, audible proof of paternal care. They'd argue about everything – he'd call her a bitch, she'd call him a fucking arsehole, keys would go flying – but on this point they were of one resolute and unified voice: the gate did not need replacing. It served as an effective deterrent to burglars (it didn't – we were knocked over three times in ten years), and it was 'easily mended'. When I finally helped my mother move out of that house some twenty years later, I took one last look at the gate and there must have been two hundred nails in it.

Sitting on the heavily-carpeted living room floor, surrounded by heavily patterned wallpaper, I remember flicking through the accompanying box of LaserDiscs, straight past anything that seemed related to the adult world: spaghetti Westerns, the 1976 *Carrie* – and then along came *Moby-Dick*. An image of a whale-monster on the sleeve. Assuming it was a cartoon, I fed the disc into the machine. Whether I'd chosen the wrong side, or it had some mechanism for remembering where the last viewer had left off, it immediately started rolling the final battle – the volume already well up because of my mother's hearing. Gregory Peck approximating Ahab's 'terrific, loud, animal sob, like that of a heart-stricken moose' (one of the weirdest lines in the book) – the sound too much for our television speakers. Then the whale (or what seemed like several whales) arching through the water with harpoon porcupine-quills. The screaming moose now on the whale, stabbing and spitting his hate; subsequently strung up in harpoon line on the side of the whale (Peck almost drowned filming that), dead arm rocking back and forth, beckoning his fellow sailors to join him; the ramming and sinking of the *Pequod*, the implied death of the entire crew, save Ishmael.

As many a trained film critic has pointed out, these minutes are an outrageously ambitious mess – a monumental attempt to capture the impossible and, methodologically speaking, a committed approximation of the insanity it's trying to dramatise. I watched it all the way to the end credits. About ten minutes.

Years later I was lucky enough to find myself sitting around a dinner table with three of the most brilliant Melville scholars of our time – Wyn Kelley, John Bryant and Sam Otter – listening to them talk about the film's screenplay and why Ray Bradbury chose to change Melville's ending. Spoiler alert: in the novel, Melville has Ahab disappear in the whiplash-blink of an eye, caught in an unfurling harpoon line that catches him round the neck and drags him instantly overboard. He dies 'voicelessly as Turkish mutes bowstring their victim'. It is in fact the 'dusky phantom' Fedallah, leader of Ahab's personal praetorian guard and counterpart 'shadow-self', that Melville straps to the body of the whale, thereby providing the subsequent generations of scholars with enough symbolic and allegorical fodder to last several lifetimes. I forget who said this (John, I think): Bradbury's rewrite rewards Ahab with the hero's death that Melville goes out of his way to deny him. Just as Shakespeare denies Macbeth his final moment of glory, Ahab is murdered voicelessly and impotently off-stage. For Hollywood's sake – for Cold War America's sake – Bradbury felt he needed to fulfil Ahab's revenge fantasy (erasing the presence of Fedallah) – allowing him to mount his white whale, stab it repeatedly and deeply with his harpoon, scream the lines, 'from hell's heart I stab at thee; for hate's sake I spit my last breath at thee', before finally drowning him in an ecstasy of crucifixion. Or a man in bed, enjoying his own post-coital bliss.

With a pride that came with being able to fire up the machine, my five-year-old self watched the scene repeatedly over the following weeks. At least once a day, probably more than once. A kid inhaling one of the most reactionary and pornographic sex scenes in mainstream Hollywood history. Maybe my pubescent awakening had occurred on that Norfolk beach – but Bradbury's brutish rewrite of *Moby-Dick* was definitely a formative, skullfucking primer.

It also occurred to me that the Skegness panic attack had essentially rehearsed the film's aesthetic choices – my mind dredging up the overwhelming score, the specific levels of saturation. According to the cinematographer Oswald Morris, the visual effect was meant to convey the sense of a nineteenth-century whaling print – of having been filmed in 1843 when Melville was at sea – colours faded and warped over the intervening years.* After discussions with director-producer John Huston, Morris decided to experiment with sandwiching black-and-white and colour film together, with the result that each shot looks as though it's been dunked in brine and then bleached on deck. The poor-to-medium quality of our 1989 TV might also have had something to do with it, but each contrast – bird set against sky, sea against sailor – formed a muted corona, as though every foregrounded entity were a moon passing across a fading sun.

In Skegness the effect had gradually intensified, as though my unconscious was having fun with the controls on a photocopier. Maybe you could put it down to the chromo-extremism of my generation's cartoon fodder: *Teenage Mutant Ninja Turtles*, *Thunder Cats*, *Transformers*, *Ghostbusters*. I watched all of those too. But in my particular case, my brain's synaptic transmission seems to have been routed through some very identifiable cranial scar tissue – scar tissue scorched into form by an outdated media format. Maybe watching the LaserDisc *Moby-Dick* when I did constituted a partial self-lobotomy. I'm reminded of Marshall McLuhan's axiom: 'The medium is the message.' The medium, my brain, cut by LaserDisc, facilitating the channelling of specific cetacean freight. A sea passage burned into my brain through which whale corpses could now freely sport.

Over the three sessions I relayed something like this to the shrink. At the beginning of the third, I explained my financial situation. She handed over some literature – potential

* See Oswald Morris and Geoffrey Bull, *Huston, We Have a Problem: A Kaleidoscope of Filmmaking Memories* (Scarecrow Press, 2006), pp. 84–5. Philip Hoare writes well about this in *Leviathan* (p. 332).

sources of funding and payment schemes. I said I'd look into it; announced I was going to Fishguard – see if I couldn't find out what happened to the Moby-Dick model. Perhaps if I reunited some of my internal mental images with their specific counterparts in reality, it all might begin settling down.

Silence again. My companion was comfortable with the silence.

They filmed the Gregory Peck scene in Fishguard in 1954. The director Huston commissioned a model, several models, of the white whale – one was about seventy feet long and it came loose during filming. The Pembrokeshire coast was experiencing the worst weather in living memory and the seas were nearly impossible to navigate. Some of the contemporary reports say the model drifted out into the Atlantic Ocean; others say it stranded somewhere along the coast. Maybe if I found part of this foundational entity – even just went through the process of searching for it – things might shift, resolve themselves into something more existentially sustainable.

The models of Moby-Dick caused John Huston and his team a good deal of trouble. The chief aim was to avoid 'a phony-looking whale', a difficult task in 1954. They had the budget, and experimented with live-action shots of whalers hunting off the coast of Madeira (an island to the south-west of Portugal). These proved only of limited value, partly because the sperm whales off Madeira were not aggressive enough, or white enough, and partly because the Bradbury script stipulated hand-to-fin combat: Peck and the whale needed to be up in each other's grills. According to the producer Walter Mirisch, Huston at one point suggested they could sidestep the phony model problem by having a shark

'The Jaws of the Great White Whale: Part of A Model of "Moby Dick" at the London Zoo', *Illustrated London News*, Saturday, 2 April 1955.

'which would be encased in a whale suit, put into the water, and somehow or other controlled on some kind of wires.'

They went for models. Most of the close-ups were taken at Shepperton Studios (I went to school in Shepperton), a life-sized reproduction of a sperm whale's head proving particularly productive in this regard. This is the model that bites Ahab's lowered boat in half with him still in it: forty foot long and thirty-two foot high, made of rubber and plastic. After filming, it was sent to London Zoo to promote the film. They had it on display for a year in the antelope paddock. I've tried to track it down, but the trail goes cold in 1957: either too difficult to clean and destroyed, or locked away in a private collection.

This head aside (impressive though it is), Huston knew that a film of this scope and ambition really needed a wide-angle shot of a decent-sized white whale in combat with Peck. Step forward special effects man Augie Lohman, responsible for building the triceratopses in *The Lost Continent* (1951) and the giant fibreglass sea-mollusc in *The Monster that Challenged the World* (1957). The obvious man for the job, apparently.

Working with the designs of art directors Ralph Brinton and Stephen Grimes, Lohman was faced with the task of selecting building materials that would be robust enough to withstand rough seas, have a Hollywood superstar scramble around on its back – and remain afloat. According to a syndicated story that was doing the rounds between October and November 1954, the whale consisted of a wooden framework 'covered with asbestos, rubber and plaster'; it contained 'about eighty drums full of compressed air to keep it afloat',

'MOBY DICK HIMSELF TAKES SHAPE: The monstrous leviathan, with whom Ahab and the whalers waged a feud to the death, is being constructed in Fishguard Harbour for the film. Workmen are seen here putting on the outer skin. This consists of rubber sheeting stretched over a wooden framework. Inside the body are tanks of compressed air.'

along with 'a hydraulic mechanism to simulate realistic movement'. It took three and a half weeks to construct and a day to launch.* There are conflicting reports regarding its precise scale. Several sources, including Oswald Morris's memoir, claim that no complete model was ever built, and while that is true, such comments belie the scale and grandeur of the thing they did eventually put together. One photograph, published in the *Sphere* in October of 1954, shows most of a more than

* '12-ton Model Whale Hazard to Ships', *Portsmouth Evening News*, Friday, 29 October 1954.

life-sized whale taking shape in Fishguard Harbour. As the caption of the image confirms, the frame of the whale was wood, and those are large sheets of rubber being draped and then stretched over the main carapace. It was seventy-five feet long and weighed twelve tonnes.

In Chapter 67 of *Moby-Dick*, 'Cutting In', Ishmael details the process of 'flensing' or butchering a whale carcass; that is, stripping it of its blubber so that it can be processed into oil. Through a series of precise cuts and fastenings – and with the ship serving as counterweight – 'the blubber in one strip uniformly peels off along the line called the "scarf"'. Judging by the photo, the construction of Moby-Dick in Fishguard Harbour must surely be one of the singular instances of a 'reverse flensing', or re-insulation of a whale. That sheet of rubber-blubber nearest the head has indeed been arranged as a sort of cetacean scarf, making me think they were taking directions from the book and working backwards.

'Film Whale Walks Out' was the headline in *The Times* on 30 October 1954. For the next two months, national and local dailies kept the British public up to speed on the hunt for the missing Moby-Dick, adrift somewhere in the Western Approaches: '"Moby-Dick" Loose in Irish Sea', '12-ton film "whale" adrift off Welsh coast'; '"Moby-Dick" not yet sighted'. The crew had been towing the model out to a filming location just off Strumble Head in rough seas when the towrope snapped. Spotting the danger, and risking life and limb, the assistant director Kevin McClory jumped onto the back of the whale 'with the courage worthy of Ahab himself',* quickly fastened another line, and jumped back.

* As reported in the *Portsmouth Evening News*, 29 October 1954.

McClory's knot didn't hold either, and in the next moment the whale was loose again, quickly drifting out of sight in the strong current. A tugboat gave chase, but in vain, and soon after the search was called off.

The production company then made two sheepish phone calls: one to their insurance company Lloyds of London informing them of the loss, and the other to the coastguard, telling them of the potential danger to local shipping.* The coastguard dispatched a vessel, and the RAF instructed one of their flying boats, on exercise over Cardigan Bay, to keep a careful lookout.† The Royal Navy issued a statement saying that it had warned all ships by wireless. After three days, the search was abandoned: 'No news of "Moby Dick" the seventy-five-foot, twelve-ton artificial whale adrift somewhere off the south-west coast of Britain, had been received this morning,' Naval HQ at Plymouth told reporters.‡ Sailing a few miles off Fishguard, a steamer had reported a single 'suspicious radar echo', but that was it. According to the *Portsmouth Evening News*, Moby-Dick must have been 'drifting out towards the Atlantic'.

---

* Melville wrote a poem about just such a shipping hazard in his little-known 1888 collection *John Marr and Other Sailors*: 'Drifted, drifted, night by night,/ Craft that never shows a light;/ Nor ever, to prevent worse knell,/ Tolls in fog the warning bell./ From collision never shrinking,/ Drive what may through darksome smother;/ Saturate, but never sinking,/ Fatal only to the other!/ Deadlier than the sunken reef/ Since still the snare it shifteth,/ Torpid in dumb ambuscade/ Waylayingly it drifteth.' Not his best work, admittedly.

† '12-ton Film Whale adrift off Welsh Coast', *Western Mail*, 30 October 1954.

‡ '"Moby Dick" not yet sighted', *Sunderland Daily Echo and Shipping Gazette*, 30 October 1954.

And 'drifting out towards the Atlantic' has since been the consensus. In his memoir, Oswald Morris rounds off the anecdote by wondering what 'shipping in mid-Atlantic made of our twenty-five-foot-high Moby-Dick as it bore down on them at a rate of knots'.*

Never been convinced by this. Hear me: in Cardigan Bay, the water tends to move in an elliptical tidal motion that circulates back in on itself – *an ideal current to strand a model whale*. Other newspapers have the model 'drifting north', along the coastline.† In the *Birmingham Daily Post*, one article headline very reasonably suggests that '"Moby-Dick" May Be Washed Ashore': 'Moby Dick, it is believed, will float up and down the Irish Sea and possibly be washed ashore somewhere near Aberystwyth.'‡

I'd do a bit of asking around. Perhaps someone had chanced upon a piece of timber or a drum full of compressed air. At the very least it would be a few days on the Welsh Coast. I'd bring a tent and a pair of binoculars, have a leisurely drive from Fishguard to Aberystwyth, and spend a day or three exploring the flotsam and jetsam. A part of the coast I didn't know so well – that everyone keeps saying is one of the most beautiful places in Britain. Alicia had a work deadline, so this'd be a bit of self-care: new associations, mental rewiring.

Passing Bristol, I decided to stop in at a service station for a coffee – opened up my Redman (I never go on the road

---

* *Huston*, op. cit., p. 88.
† *Western Mail*, 30 October 1954.
‡ *Birmingham Daily Post*, 1 November 1954.

without him, cumbersome though he is to carry). Another dead whale badgering me for attention. In St Mary Redcliffe's church: a bone brought back by John Cabot in the late fifteenth century and transformed into the sacred rib of the Dun Cow, the mythical creature with the infinite supply of milk. I parked at Temple Meads station, walked the short distance to the church. The rib sits on its own corbel in the Chapel of St John. I lit a candle and asked the whale-cow to bless my journey.

A church volunteer approached, younger than me. He was friendly and informed – explained that regional variations of the Dun Cow myth had fused over time. The animal belonged to a Warwickshire giant; it had escaped and started wreaking havoc on the local community after being driven mad by the over-milking of a witch. It was eventually apprehended and slaughtered by Guy, Earl of Warwick.* Before the St Mary's bone became associated with the cow, it was widely said to be part of Guy himself. Others say it is actually part of St John. The volunteer smiled at the thought.

---

* Another whalebone rib is still on display in Warwick Castle. See Charles Hardwick, *Traditions, Superstitions and Folklore (chiefly Lancashire and the North of England)*, 1872. Qtd. Redman (213). See also the bone at Dun Cow Rib Farm in Wittingham, Lancs, a story involving another witch: 'there was once on the moors an old dun cow of great size, which had no owner but gave milk freely to all comers. An old witch once took a riddle instead of a pail, and the cow, mortified at being unable to fill it, died. The people much regretted its loss and preserved its ribs for a memorial.' *A History of the County of Lancaster*, Volume 7, p. 70. Originally published by Victoria County History, London, 1912. The rib – a whale's or a Bronze-Age aurochs's – still hangs above one of the entryways.

I asked if he'd heard the story of Letitia Bonaparte and the shoulder bone of St John. He glanced over my shoulder at another tourist. 'Another time', I said.

'No, no – please.'

'No, it's a busy time.'

'I'd like to hear the story.' A kindly firmness to his voice. He waved at a colleague with the tips of his fingers – eyebrows raised in broad Anglican friendliness. He returned to me. I opened my mouth. This swam out:

> In 1805, a young, and by all accounts very handsome, priest arrived at the Palace of St-Cloud just outside of Paris and demanded to speak to the Mother of His Imperial Majesty the Emperor. He claimed that he had recently visited Syria and had managed to acquire a famous relic, the shoulder bone of St John the Baptist. Unfortunately, he hadn't had enough money to get to Paris and present this to Her Majesty in person, and so had been forced to pawn it to a Grecian bishop in Montenegria for two hundred Louis d'ors (22–23 carat gold coins). Anxious that the relic be delivered safely to her collection, Letitia immediately gave him the money, and advanced him a further hundred for his troubles. Within three months he had returned with the shoulder bone of St John and, after requesting a further not insubstantial sum for other costs incurred, delivered the relic to its delighted new owner. Meanwhile, Letitia's ladies in waiting had become suspicious of the charming young priest who kept receiving gold from their mistress and suggested that his story be checked out. He was soon revealed as a fraud and a French deserter, and on

interrogation admitted that he had sold the Emperor's mother the jawbone of a whale.*

'Won't find that in your Redman,' I concluded, indicating the book under my armpit.

My Anglican friend looked thoughtful. 'Something similar happened to my grandparents,' he said. 'Some criminals recently called up and insisted they'd won a new patio – fossil buff stone, free installation. Stole five thousand pounds in the end. It happened last week – all a little close to the bone.'

I was back in the car. It was raining hard. I was driving under water. At the very least I'd look out from Strumble Head lighthouse, where the Peck-Dick battle had unfolded. I'd get a visual, connect up my brain to one or two tangible geographical features, tether it to solid ground.

A four-hour drive to Lower Town and the Ship Inn. I spent the evening drinking, listening to the rain. First on my own, and then with Gavin and Andy, both in their sixties. Andy's nickname was 'Bonfire', on account of his pipe smoking, and his complaints about the ban that came in over a decade ago. They'd been sitting, largely in silence, at

---

* According to the main source, written by the contemporary anti-Catholic propagandist Louis Stewarton, 'Madame Letitia did not resign without tears the relic he had sold her; and there is reason to believe, that many other pieces of her collections, worshipped by her as remains of saints, are equally genuine as this shoulder bone of St John'; in *The Secret History of the Court and Cabinet of St Cloud: In a Series of Letters from a Resident in Paris to a Nobleman in London, Written During the Months of August, September and October 1805, Volume 1* (London: J. Murray, 1806), pp. 232–44.

the table next to me. A packet of tobacco on the table – a particularly vicious photograph of some dead lungs where the Marlboro logo used to be. Two beers in and Gavin asked me what I was doing in town. Alcohol in my sails, I told them my plan. They looked at me thoughtfully. Bonfire said they'd never heard of anyone finding anything, explained they'd lived there most of their lives, had seen the model being built with their own eyes – would have heard about it if Moby-Dick had come ashore. He called to the bartender, Mike. Mike agreed that they would have heard about it. I must have looked crestfallen. Gavin took pity: 'Buy us a round and we'll tell you some stories.'

The evening progressed. They'd pop outside for a smoke, agree on a yarn, and then spin it on their return: fishing stories, mainly, but also drinks with Richard Burton (he'd filmed *Under Milk Wood* here in 1972). Mike the bartender was Peck's bastard child.

An hour and a half into the session, and Gavin asked if I was staying at the Bay Hotel.

I wasn't.

'You know about it, though?'

I didn't.

They looked at each other. It's where they'd all stayed. They completely took it over – hundreds of them. Stars, film crew, technicians. The works. They renovated the place especially for them.

I asked if it was expensive. Not too bad. A three-star, not far – just up beyond Goodwick. And it's low season. I'd probably get a deal. Worth a look. We said farewell. I was too drunk to drive. An hour later I was at reception asking where Gregory Peck had stayed, and whether I could have the room. 'Room 123 usually requires advance booking,' said the young man behind the desk. I asked if they had anything overlooking the harbour. They had a twin room. With a view? Not of the harbour. Sixty including breakfast. Seemed reasonable. As I filled out the forms, I asked if they'd all stayed here, the film crew and actors. He pointed at a photograph on the wall.

Peck on the right in the magnificent white coat (an allusion to the whale?) listening to Leo Genn (Starbuck); Huston in the middle and flat cap talking to the assistant director Jack Martin; Richard Basehart (Ishmael, he looks like David Bowie here) raising a glass to the ship's carpenter Noel Purcell. The man leaning against the bar and looking lost is Johnny 'King' Kelly, apparently, a stunt man who didn't get a credit, in spite of his title. Somewhere in here you can also find electrician Len Prout, celebrated cameraman Jack Culver and make-up artist Len Garde. An accompanying caption:

> The Fishguard Bay Hotel has been taken over by the film company making *Moby-Dick*. The hotel has been closed for the past two years, and has been entirely re-furnished including new bathroom fittings, new bedroom suites, fitted carpets, etc., at a cost of £15,000. All the stars of the film, including 125 film technicians, carpenters, stunt men and the crew of the boat *Pequod* are living as one big happy family. Everybody mixes quite freely and it is quite a common sight to see Gregory Peck and the stars having a drink in their newly opened bar along with carpenters and electricians. The film will be the most expensive ever made in this country, and the estimated cost is around one million pounds.

'So the hotel became a kind of *Pequod*,' I said, with slurred profundity.

The curtains to my room were drawn, and I opened them to reveal an ivy-covered wall. A dry T-shirt and I went back down to the recently refurbished bar, asked for a gin martini, the most Hollywood drink I could think of. Two pensioners were sitting by a window in silence, sipping half pints of bitter. I tried to reconstruct the scene – where they'd all been standing – the movie-star repartee. What it must have been like to live at a time when the threat of extinction was only nuclear.

A decent hangover the following morning, the kind that tells you you're not young anymore. At least midday before

I could legally drive again. The rain wasn't going to stop. The views from the hotel breakfast room are stunning, apparently. On the way out I got chatting to the receptionist again. He told me that it'd be better if I didn't drive to Strumble Head today because I wouldn't be able to see anything. I didn't meet a single car on the fifteen-minute drive. At a viewpoint marked on the map, the horizon consisted of a blasted heather clump in my immediate foreground and otherwise a pure whiteout. I just sat there, staring out.

A text message from Sandy. How was I doing?

Not great Sandy: experiencing the world as a palimpsest of dead and stranded whales atm.

There have been a lot of them recently. Did you visit the new arrivals?

Yes.

She waited for me to elaborate.

Didn't feel like me doing the visiting; dead whales in me visiting the dead whales. Does that make sense? Like I've become a cetacean spirit medium.

A lengthy pause before the reply.

Where are you?

Wales.

ha.

Fishguard.

Lol

?

A tremor of fear in my chest.

Maybe I'm being paranoid [I wrote back], maybe I'm seeing patterns that aren't there – but why are there so many of them all of a sudden?

Relations playing up?

Panic attack in Skegness when I visited them.

A pause.

She said that the offer to help track down Blue Hair was still there. She had someone in mind. Nice guy, grew up on the east coast and knows it really well. Just in case I was interested.

Most likely it was rooting around in your garden Sandy that got me into this mess in the first place. I didn't write that. I put the phone in the glove compartment. In the mist the faintest outline – an arcing severed tail among the boiling vapours. She was probably right. The dead whale's jaw – a gigantic punctuation mark I'd just let passively hang over my life – for years. I'd start doing some proper work. This was all just an evasion. Who was she? Had I misremembered her – the tone of her voice? I was so young. Had she been suffering from a similar affliction? Is that why she took the jaw? To try and shut them up – silence them? I'd always assumed her goodness. Maybe she'd handed over some kind of burden-curse.

4

Torquay

I began researching the week in question: the comet, the whale, the jaw and its beautiful scavenger. Something must have made these things line up. The comet tattoo: a celestial event that made an indelible impression on her. I found out that my morning in the dunes coincided pretty much exactly with the moment when Marshall Applewhite and the 39 finally beamed up to the spacecraft they claimed was tailing the Comet. In this instance 'beamed up' means 'committed mass suicide'. Applewhite had been the leader of the Heaven's Gate cult. He was adamant – it had been the 'Last Chance To Evacuate Earth Before It's Recycled'.

Comet woman had been a follower – performed a ritualistic defacement of the Christian hell-mouth. A second 'Harrowing of Hell': a vandalisation of the door through

Last Judgment fresco, by Giacomo Rossignolo, *c.*1555.

which the disobedient, sinful and independently minded have traditionally passed and been condemned to eternal torment. A Boudican witch-pirate destroying the infrastructure of traditional, patriarchal morality.

Hers was an attempt to destroy the Hell Gate – so that Heaven's Gate might open elsewhere: the shutting of one gateway to open another. After all, the jaws of the whale as hell-mouth is a long-established tradition: in *The Exeter Book*, a tenth-century anthology of Anglo-Saxon poetry held in the vaults of Exeter Cathedral, there's a poem called 'The Whale'. It speaks of Fastitocalon, a creature that disguises itself as an island, tempts sailors to camp on its back, and then drowns them by descending back into the waves. The second part of the poem describes the 'sweet stench' of 'Hell's ornate doors' – the scent emitted by a whale's mouth to attract sinners. Scything off the jaw was not only a refusal of aesthetic temptation, but also an effort to refigure the parameters of Christian eschatology itself. Applewhite had a partner, Bonnie Nettles, the co-founder of the movement, who exited her 'vehicle-body' before her time on 19 June 1985. This happened to be eleven days before my birth on 1 July 1985. 11+1 – an apostolic figure if ever there was one; 111 – a new trinity? Members of Heaven's Gate were adamant that Nettles continued to help them from the 'Next Level': had I been aiding the holy ghost of Bonnie Nettles? Was I her reincarnated?

No, I wasn't. But the patterns and connections were establishing themselves all over the place. And for this moment of personal disintegration to be met by the largest mass stranding of whales in recorded British history – surely that meant something? Surely it all connected up to a grander scheme? I just couldn't quite see it yet.

Alicia listened carefully, sympathetically to all this and, maybe with a touch of urgency in her voice, suggested that it was perhaps time to come to some sort of understanding or arrangement with 'this side of the family'. Continue with the psychotherapy? We could find the money. Take Sandy up on her offer? Maybe both? Looking back, I puzzle at her forbearance: encouraging me to come to terms with something that might easily have been construed from a certain angle as an extended teenage crush, a state of arrested development. Never look directly at such things. Just be grateful. If the whales were indeed guiding me along, they'd been breathtakingly generous in their choice of human companion for me.

*

Mick had grown up in Norwich – was 'definitely worth talking to', according to Sandy. Knew people who knew people. Knew a thing or two himself. On Saturdays he stood sentry at a local flea-market stall. I decided to pay him a visit. I put the keys in the ignition; on came the radio and a stranded whale had chosen the song: 'God Only Knows' by the Beach Boys. As I drove, my head filled with the various ailments, foibles – one might describe them as 'beachings' – of the three Beach Boy brothers, Brian, Dennis and Carl Wilson: the abusive father, Brian's depression, Carl's lung cancer, Dennis's dalliance with the Manson family cult. I pulled over, located the song on my phone, and played the *Live at Knebworth* version on loop until I got to Torquay. Nineteen eighty. Carl's voice slightly older and impossibly lovely in this recording. One of the band members, I think it's Brian, says 'Very pretty, Carl' as the song winds down after the in-the-round outro.

A man in a fluorescent jacket guided me into a Devon field, another into a space. I switched off my engine and left the music playing; watched the elevated and brand-new wing mirror of a next-door SUV retract itself magisterially away from my face. This was big for a regional flea – tables lined up as far as the eye could see. I moved down the avenues, in my Sub-Sub element. Sandy had been slightly hazy on the exact location of the stall, said I'd recognise it when I saw it.

Fragments of First World War shells and empty cartridges on one table. 'Vote Leave' bunting strung out across its entire length. The centrepiece ('not for sale') was a slice of tree stump – jagged pieces of iron embedded deep in the wood. I asked where these pieces had come from. Slightly on the defensive (the French and Belgian governments almost certainly have an embargo on the export of this kind of material), the man explained that his uncle owned a farm not far from Vimy Ridge.

'How much for this bit?' – a piece of shell about three inches long. Tenner for the larger shrapnel. And bullets? Four for the cartridges. I pointed at a grenade fragment.

'Tank shell. That's also a tenner.'

Over the years I'd assembled some bits and pieces about stranded whales during the world wars. One in particular now swam to mind (inevitably by a man named W. Beach Thomas). He'd gone so far as to explicitly figure immobilised tanks as stranded whales, vanquished leviathans:

> GLORIOUS DEEDS OF LAND SHIPS IN THE BATTLE OF ARRAS. MONSTER'S SUCCESSES. FROM W. BEACH THOMAS.
>
> War Correspondent's Headquarters, France. Wednesday. —One of the undoubted successes of the Battle of Arras was the Tanks. The Germans claim to have destroyed twelve, and it is true that a few of these stranded whales lie about the battlefield, but whatever proportion have or have not been hit, they have saved hundreds, perhaps thousands, of lives.
>
> *Daily Mirror*, Thursday, 19 April 1917.*

A nice thank-you present for Sandy – another piece for her collection. The man explained that I'd get a certificate

* See also 'DEVASTATION IN APRILLA. COUNTRYSIDE RESEMBLES SOMME BATTLEFIELD. WRECKED TANKS LIKE STRANDED WHALES.' *Scotsman*, Tuesday, 30 May 1944.

of authenticity, signed by a local historian. I asked who that was. 'Alan' – he reached for another word – 'Michel'. A slight French lilt given to this second name. I handed over the tenner and walked away with my fragment of war whale.

A table of books. A set of almost-new Nigel Slater. The well-spoken woman behind the table told me they were two pounds each. A babe-in-arms played with the 'Remain' badge on her lapel. Feeling the need to justify her presence there, the woman told me they were having a clear-out, even though it was obvious to both of us the only reason you'd get rid of your Slaters was divorce, redundancy or death. I picked out a copy of *Real Fast Puddings*, fished a two-pound coin out of my pocket and rubbed something off the Queen's face as I handed it over.

I was stopped for a quick coffee at a nearby stall, slouched on a deck chair reading about pears with soft cheeses, blackberries in Barolo, a grape tart, when I looked up from the Slater and did a double take on a set of sperm whale teeth upright on a table across the way. Just sitting there among a series of Royal Doulton ceramic dogs, a Denby tea service, an old pair of military binoculars. I stared for a few seconds, made sure they were real, closed the book, and wandered over.

The man standing behind the stall was about 5'6", stocky, probably in his fifties. In spite of the heat he wore a battered leather jacket and his hair was dyed jet-black. Neck the colour of bing cherries and a Vote Leave badge pinned to his chest. I asked if I could take a closer look at one of the teeth. He handed one over. The label on the side said 'Scrimshaw: £60'. A mermaid etched on one side; a British man o' war on the other, the Union Jack raised and fluttering in the breeze.

A fake. You can tell by the untextured cross section of the resin base, if not the weight. I asked if it was a copy.

He looked at me. He had the clear hazel eyes of someone half his age. 'If he was real I'd be selling him for more than £60.' This was a high-quality replica. A celebration of the great British whale.

Did his name happen to be Mick?

He told me I was Peter; that Sandy had mentioned I'd be popping down.

Handing back the tooth, I asked if he ever got real ones in.

Rare to come across them nowadays – the law and all. Reproductions from China, mostly. Good ones, mind, no plastic. He used to have a stall up on the Portobello Road – before him and the wife moved down here. They're popular: people put them in their bathrooms and hallways. Memento of the good old days. The way things used to be.

He saw the brown paper bag I was carrying, asked me what I'd bought; seemed impressed (maybe even a little taken aback) when I told him about my present for Sandy, my fragment of stranded war whale.

He turned it over in his hand. Apparently I'd thrown down something of a gauntlet. For the next hour or so, we exchanged anecdotes, fragments, esoterica – a distinctively competitive edge to proceedings, each working out what layer of subcultural involvement we were respectively dealing with. In spite of interpretive differences and divergent political affiliations, the conversation was reassuringly collegial, pedantic. Special subject: World War Whales, 1914–45. The whales rallied round, dredging my knowledge and supplying me with ample material to keep up. Mick began with the forgotten heroics and sacrifices of 'Mr Whale' who,[*] he said, really ought to be posthumously recognised by the nation. Whale oil's glycerine content was the vital component in the manufacture of high explosives during both wars. Keen to stress my credentials, I came back with a pleasingly precise anecdote about an organisation called the National Salvage Council, which during the First World War quickly descended on any new arrival, armed with knives and hatchets, in order to extract said oil as quickly as possible: it knew more than anyone else that whales translated directly into bombs.[†] Our conversation soon turned into a form of enjoyable brinkmanship, taking each other on trust for the mainly misremembered anecdotes concerning particular war whale victims, variously blown up by mines, targeted by returning bombers, or torpedoed as enemy submarines.

Mick caught me off guard by suddenly tacking to recent

---

[*] See for example 'WHALES ON WAR WORK. Whales don't wear battle dress, but they're playing a big part in this war ... Many of the lads in the Invasion Forces owe a debt of gratitude to Mr Whale, without always knowing it. From whale oil, various ointments and fibre dressings are made. Without glycerine it would be hard to manufacture high explosive, and whale oil is the chief source of glycerine. *Good Morning*, Thursday, 3 August 1944.

[†] See also 'WHALES AND THE WAR. A whale was recently washed up on the low-lying sandy beach at the Suffolk coast. Its weight was about 14 tons, and the carcass was claimed by the National Salvage Council, whose duty it is secure the fullest possible use of waste material for national needs ... a fatigue party at once attacked the whale with knives and hatchets and the oil was extracted in a "digester." The resulting glycerine should be sufficient to provide the propellant for some 130 eighteen-pounder shells, and the bones will go to increase the supply of manures so urgently needed for the land.' *Southern Reporter*, Thursday, 30 May 1918.

history: had I heard last year's report by government scientists on the Royal Navy's killing nineteen pilot whales on the north coast of Scotland in 2011? I confessed I hadn't. Seventy long-finned pilot whales – Edwards – had swum into the Kyle of Durness, apparently, one of the world's largest live bombing ranges.* He let the anecdote settle. I'd been vanquished. He pointed at my piece of tank and reaffirmed how thoughtful a present it was. I bristled slightly at the condescension, but felt a lightness on my chest. There were others like me: I needed to hold on to that. I wasn't alone in this. He said his business partner was coming over for drinks that Tuesday evening. Between them, they'd put their heads together and see if they couldn't help me out.

Tuesday afternoon, I kept the appointment. Mick opened the door to me, explaining that Ron was stuck on the M5 but would be with us soon. Nice place. First-floor flat in a substantial Victorian house. Clean and tidy. I was led into a high-ceilinged kitchen-diner. A framed poster of Elvis Presley hung above a boarded-up fireplace. Late sixties full black leather. Mick saw me looking. 'Here', showing me to the fridge-freezer. A magnet with a photograph of him as a younger man. Full Elvis regalia, clutching a pencil microphone, face emphatically contorted as he sang 'Love Me Tender'. Girls 'falling over themselves, they were'.

I smiled politely; asked if he still performed. Not for a while – his back had gone a couple of years ago. On stage? No, picking up a shopping bag. Put an end to it all. If you

* 'ROYAL NAVY BOMB EXPLOSIONS CAUSED MASS WHALE DEATHS, REPORT CONCLUDES,' *Guardian*, Wednesday, 24 June 2015.

watch, Elvis gets himself into all sorts of positions. He raised an arm in half-demonstration, told me he'd stopped doing his stretches. Half an hour a day he used to do his stretches. Kept him going well past his sell-by. 'Always do your stretches. Keeps you young.' He looked down at his paunch, muttered something.

'A great photograph', I said, trying to maintain the glory-days tone.

He said he had a box or two I'd be interested in, disappeared through a door and walked down some steps to the basement. In his absence I scanned the room: a display cabinet collection of novelty mugs, a novelty clock, the face of which was a picture of Mick and his ex-wife looking happy. As I walked over to take a closer look, a very loud doorbell rang – one long blast, followed by rapid-fire shorter rings. Mick shouted up to say he was coming; emerged carrying a large brown box labelled 'TEETH'. Bit of a character, Ronnie, but I'd like him. 'A big boy. Passionate about his whales.'

The man who entered the room had to mind his giant balding head on the doorframe. He stuck out his paw and stalked over to me. An impeccably groomed anchor-shaped goatee – faded purple dye – starting just beneath the middle of his bottom lip and spreading out along the ridges of his jaws-bones. He wore a band T-shirt: 'The Business: Strength Thru Oi!' emblazoned across it. I initially did a double take and read this as 'Strength Thru Oil' (when I looked up the band I discovered that two of its long-standing members were called Steve Whale and John Fisher). A well-moisturised handshake, Ron apologised for being late – the M5 was 'fucked'. Mick said he was just about to show me their collection of teeth, and that I knew a thing or two.

'The box of false teeth.' Ron said this affectionately. He reached into the box and pulled out a *faux*-sperm whale tooth: a portrait of Nelson framed by a decorative wreath of laurel leaves. They'd started ordering these from a factory in Hebei in the mid-2000s. You send them the design and they do the rest. In fact, they'd just put in an order for a commemorative Brexit tooth to mark 'the nation's imminent rescue and re-floatation'.

I avoided this and asked if I could take a look; walked around the table and stared into the box of whale teeth. Decent resin reproductions. About thirty of them jumbled together. Nelsons, mainly, the occasional pair of tits. They looked good, sold for as much as $150 to the right buyer. Most of the trade was online now. 'All in the description' – use a legally exonerating word like '*faux*', and then double down on 'rare antique scrimshaw'. Usually did the trick.

Ron wheeled over a large suitcase. Stickers all over it: 'Handle with care', 'Fragile', an assortment of airport baggage handling labels. He placed it on the table, unzipped the top, and opened it away from him so that only he could see the contents. He started unpacking, arranging all sorts of bubble-wrapped odds and ends on the carpet including a framed and signed photograph of the former Manchester United manager Ron Atkinson. 'There's Ronnie,' he said as he laid the bubble-wrapped frame on the floor.

He found what he was looking for. It was wrapped in a tea towel. With a flourish, he revealed another tooth and handed it over. The Duke of Wellington. On the base there were obvious signs of the factory moulding (someone hadn't sanded it down properly). I turned the object over in my hand; asked if this was another copy.

The two men looked at one another. Ron put his hands up and grinned. A little test they liked to do. See if the person they were dealing with was kosher. He then retrieved a Tupperware box and handed it over. I'd be interested in these, for sure. Nestled in another tea towel, two genuine teeth, one in perfect condition, one slightly damaged, probably during the process of extraction. Then another bubble-wrapped item. A bone: the sun-bleached lower vertebra of a fin whale, spinous and vertebral processes snapped off, but otherwise perfect.

I examined them; asked where they'd come from. Ron said we ought to talk about me first. Sandy had given them an outline – who I was trying to track down.

Not sure yet whether I wanted to track her down – I'd been experiencing one or two problems recently. The whale thing had got out of control – wanted to know whether she suffered from a similar ailment, whether she'd sawn off the jaw in order to try and silence them, 'turn them off', shut them up. Or whether I'd been cursed that day – there'd been a comet in the sky. Had she transferred an ancient hex? I was fairly recently married – wanted to make that work.

To my surprise, Mick said I was in good company – 'massive year for them' – something was definitely in the air. They'd noticed it too. 'Just because you're paranoid, don't mean they're not after you.' Dozens turned up in January and February and, at least by their reckoning, this certainly augured something. We were living through a historic chapter in our nation's story. Ron began telling me about 'the old rivalry': the British whale and the French elephant were at it again. History repeating itself. The whales that had turned up on the coast at the beginning of the year were a warning to us all. Blindingly obvious if you thought about it: the stranded whale embodied a vision of what would happen to Britain, happen to us all, if we continued to exist under the yoke of Brussels' domination. With rising passion in his voice, Ron explained that if we weren't careful – didn't vote leave – then the British whale would continue to expire, suffocate. 'It needs to get off its arse and sharpish.'

Something of a paraphrase of what they said now: but no wonder the stranded whale was currently exerting its presence in my life. Mick and Ron found it genuinely moving that they should show up in such numbers at this pivotal juncture. They were a sign of what would become of us all if we didn't start thrashing around a bit. Under the current political regime, Britain's leviathanic sovereignty lay dying. The defeat of the British whale by the continental European elephant wasn't inevitable, but the vote needed to work out. Get the next few weeks right and Mr Whale would swim again, meet the elephant-behemoth in battle again: reassert a pride and identity that was so sorely lacking. Brexit was the rescue party, the tide of a new prosperity. Or at the very least, the beginning of the rescue party, a moment of rehabilitation and resurrection.

Such an argument certainly has its antecedents.

Extract from David Armitage's Caird Lecture, which he delivered at the National Maritime Museum in 2006:

> Behemoth and leviathan, the greatest beasts of land and sea respectively, were sometimes identified in biblical commentary with a crocodile and a bull but, more often, with the elephant and the whale. These two mighty creatures, the one predominant terrestrially and

the other oceanically, later became metaphors for power over the land and hegemony over the sea. It was thus in one sense ironic that Thomas Hobbes chose Leviathan, the great sea-monster, as the image of sovereign authority in the territorial state. More conventionally, Napoleon Bonaparte compared France and Britain to the two great monsters, the French elephant representing Europe's greatest land-power, the British whale its – and soon the world's – greatest power by sea.

The diverse political fortunes of the Great British whale:

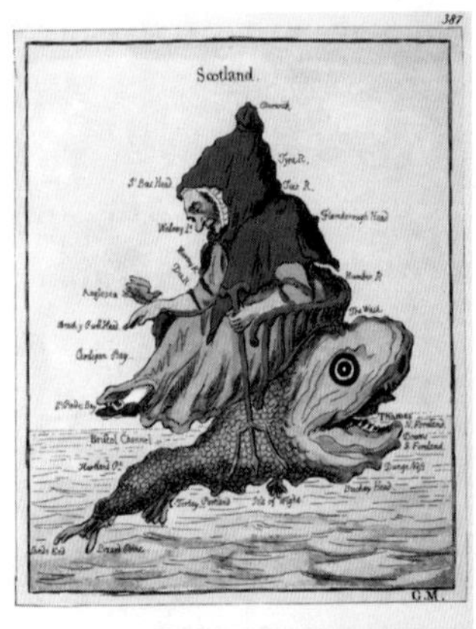

'Britannia', an etching by James Gillray, published in London by Hannah Humphrey in 1791. As the first French constitution was passed by the Revolutionary National Assembly, Prime Minister William Pitt declared Britain would remain neutral in any war against France. Britannia is meekly allowing the dove of peace to land on her right hand, while riding the washed-up, rotting remains of British monarchical or Hobbesian 'Leviathanic' power. In the 'Extracts' section of *Moby-Dick*, Melville includes a quotation by Edmund Burke that figures another European power in precisely these terms – 'Spain—a great whale stranded on the shores of Europe' – but more often than not it was England that received this satirical treatment.

Dominique Vivant's 1794 engraving 'Le Phallus Phénoménal' – perhaps the most extraordinary representation of stranded British power. The beached whale is exchanged for its symbolic referent: a gigantic guillotined phallus.* Vivant or the Baron de Denon was a French artist, writer, diplomat, author, anatomist, archaeologist, original director of the Louvre – and pornographer. This is a satire of big dicks – the marooning of monarchical sovereignty and the assembled host of revolutionary oglers and scavengers. Following the example of France, the British leviathan's inevitable fate: the (British and French) monarchies' awful end. The critique is also self-reflexive: Vivant etched this during the Reign of Terror; Robespierre was guillotined in July 1794.

* Vivant is also parodying Jacob Matham's endlessly reproduced 1598 engraving (after Hendrik Goltzius) of a sperm whale stranded on the beach at Katwyck. See p. 16.

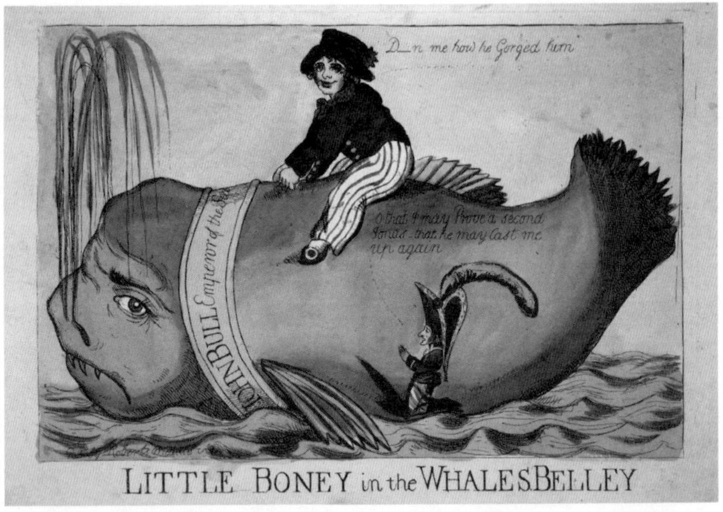

LITTLE BONEY IN THE WHALESBELLEY. (Attributed to Roberts, 1803). A more confident British cartoon, boasting of sea-going supremacy. Mythic personification of Britain, John Bull, rides a tamed Jonah's whale (suffering from Napoleonic indigestion – the Emperor feminised as ambergris). The whale as Bull's giant dick. The collar (or cock ring-cum-bondage collar) says: 'John Bull, Emperor of the Sea'. Napoleon's praying arms casting the shadow of a phallus in front of him; his hat a flaccid phallus drooping behind him: 'O that I may prove a second Jonah – that he may cast me up again.' 'Little Boney' positioned as the whale's dick, pleading to be ejaculated onto dry land; Bull riding up top in a pornographic assertion of dominance and same-sex desire. His caption: 'Damn me how he Gorged him.'

A YORK ADDRESS TO THE WHALE. CAUGHT LATELY OFF GRAVESEND: 'O Mighty Monster of the Deep, continue to attract the attention of John Bull, bend his mind solely towards thee, for in that is my only hope – fascinated by thy powerful attractions he may perhaps forget the honour of a P---c.' (5 April 1809, published by Ths Tegg, No. 111 Cheapside). The Duke of York pleads with a stranded whale (recently caught at the mouth of the Thames and put on public display) to divert the British public's attention from his embarrassing court case. George III's second and favourite son, commander-in-chief of the army, in trouble over the sale of military commissions by his former mistress, Mrs Mary Anne Clarke. York as a stranded whale: his hat positioned as a tail.*

---

*This particular whale, stranded near the mouth of the River Thames, was the subject of an extraordinary legal battle concerning ownership. 'A contest has arisen between the Lord Mayor and the Admiralty with respect to the right of property in this fish. The former claims it as Conservator of the Thames; the latter consider it as a Droit: and accordingly the Marshall of the Admiralty seized it, notwithstanding the Lord Mayor's protest against such a proceeding.' *Chester Chronicle*, 7 April 1809. When the dispute was heard at the Court of King's Bench in 1809, the ingenious lawyer for the Admiralty (the Crown), one Mr Dampier, decided it was expedient

THE PRINCE OF WHALES OR THE FISHERMAN AT ANCHOR. (1 May 1812). Three years later, York's brother as an extravagant dandy whale in 'the Sea of Politics' with encroaching beaches on either side. Prince George (who'd become Regent in 1812) surrounded by mistresses and flatterers. George had angered his former Whig friends by retaining the Tory Spencer Perceval as his chief minister. Perceval is sprayed by the 'Dew of Favor', holding the prince by a golden anchor. The Whigs receive only 'the Liquor of Oblivion'. Not interested in politics, the Prince ogles his most recent conquest, Isabella, Marchioness of Hertford, while a horny merman swims erect behind her.

Listening to these men speak, the nationalist implications of their project made a strange kind of sense to me. By their account every single whale tooth or whale bone sold was a remobilisation – a symbolic rehabilitation of the British Empire. An affirmation and resuscitation of the Great British Whale. These islands were scattered with the remains of a formerly glorious body-politick – these men

---

to alter the legal status of the whale, uttering the amazing line: 'The whale is here made Defendant, my Lord.' The judge, Lord Ellenborough, ruled in his favour. From the *London Examiner*. LAW. Court of King's Bench, Monday, 27 November.

were trying to animate this corpus anew. Bring it back to life in yet another retelling of the Isis/Osiris myth. And what better way to annoy a bunch of Europhilic lefties than to start putting these fragments of former British glory back into circulation? *Who did these fucking Remoaners think they were, rallying to the defence of the global elites? Che Guevara?* Ron said I might be interested in seeing a particular piece they'd just shifted, sold to a Japanese buyer on the dark web. Ron unzipped the bag and looked up at me with another of his goatee smiles.

What I saw stopped me. A foot of sperm whale rostrum, starting from the tip and extending back four teeth. Ron picked it up – by the teeth – bracing his arms against his belly, and put it down on the carpet. A house clearance in Andover a few years back. Used as a doorstop. I walked across the room, picked up my satchel, and got out my Redman. Andover sperm whales. No Andover sperm whales. Holding up the book: 'Not in here'. They nodded, then Ron's expression softened, became reflective. The faint smell of stale oil rot rose up as he spoke.

More easily transported in sections like this. You don't want to be caught sawing off a whale jaw.

'I'll say this for her: your girl had courage.'

'Have you …?' I trailed off.

He told me to ask away; a pleasure to share. Most of their trade was antique, but it wasn't unknown for Ron to occasionally go a-harvesting. It had started on holiday as a teenager. Near Margate. Sperm whale had washed up on a Friday evening – he'd spent all night messing about with it, standing on it and what have you. 'Fuck all else to do – pigs turned up and told us to piss off.' He went back later to have another look, and then just began picking up bits and pieces along the way.

I asked if I could hold the section of rostrum – the first time since 1997 I'd cradled a severed jaw.

By all means.

I took it awkwardly into my arms. An amazing piece. I asked about the house clearance, who it had belonged to. An old timer, apparently. And the Japanese buyer – what did he want with it? And how much? Mick gave me one or two answers and then promptly changed the subject: said they'd better come out with it. There was someone who they were fairly sure would be able to help me. A man they'd done some business with in the past – who lives in the North.

I finished the sentence for them. Big Blue.

They seemed surprised.

I said I already knew the man (was a touch insulted by the insinuation I wouldn't have known about him). Many of us know about him. Or of him. I knew his partner too. I wasn't going anywhere near Big Blue again if I could help it – not after the last time. Mick looked at me apologetically; said they'd already reached out on my behalf. Assumed they'd be doing me a favour. I thanked them.

To be fair, he probably was the best person to ask.

Britain voted. Sandy got in touch, asked (I assumed ironically) if I was going to Ron and Mick's Brexit celebration party; how I was; if they'd given me anything useful. I said I was feeling a bit better. No, I wasn't going to a Brexit celebration party, no matter how many whales were involved. Interesting people, though. Given me a lot to think about. She suggested a catch-up, on a beach, at the very precise time of 8.45 a.m. – 'Definitely not just you – beached whales are everywhere atm.'

It was being filled up with water when I arrived: a model whale, maybe about fifteen foot in length – two tonnes when full. The start of a blazing hot summer's day, already warm by early morning. We sat as close as was reasonable, trying to overhear the instructor. Sandy told me not to feel guilty about watching from the sidelines; she's a regular donor.

An insurance underwriter, a chef, a plumber, and an IT consultant, among a group of about fifteen people all dressed in wetsuits, had paid just under a £100 each to be trained up in the art of whale rescue. An organisation called the British Divers Marine Life Rescue (BDMLR) was running the course. In 2019 alone, according to their website, they held 37 such sessions and had a total of 1,788 call-outs (1,658 seals, 94 dolphins and porpoises, and 36 whales). It's not clear how many of these creatures survived, but the organisation has undoubtedly helped a large number of stranded animals since its founding in 1988. It was formed by a group of divers and vets in response to the late-eighties phocine distemper virus epidemic that resulted in the death of thousands of common seals – with their brief gradually expanding over the years.

The instructor, an articulate and passionate man, outlined to the assembled group some key distinctions between the various animals that tend to be found on British beaches, 'not all of which necessarily need rescuing'. He spoke of a number of recent incidents in which seals, resting happily on the shore and digesting their food, had been suddenly set upon by concerned members of the public and forcibly dragged out into the water. Some of these animals had become so distressed they'd gone into cardiac arrest. So avoid those, and discourage others from interfering with seals. Whales, dolphins and porpoises on the other hand – they should never be out of water, and that's the reason everyone was here today: 'how to re-float a two-tonne (life-sized model) pilot whale.'

I asked how Edward was. 'He's well – still sprawled out on the dining room table, refusing to lift a finger.' I handed over my piece of tank shell, explained the provenance. Sandy seemed pleased.

We watched as the group waited patiently for the tide to arrive – one or two making sure the animal remained wet with the occasional bucket of water (they studiously avoided the blow-hole – suffocation by damp towel is not unheard of). Everyone was also sure to stand well away from the animal's rear, in case a flick of a tail should snap one of their well-meaning legs in half. Sandy and I had the distinct impression that this was less about whale rescue per se, and more about preventing people from doing both themselves, and the animals they were trying to save, serious injury.

As the water approached, a shout went up. Four or five of the assembled rescuers started digging a trench around the model whale so as to be able to more easily position the floatation device. Another group 'monitored for signs of injury and distress'. And in the heat of this moment, the insurance underwriter, or maybe it was the IT consultant, instinctively moved towards the head of the animal and started whispering reassurances into its ear.

Sandy asked what I'd thought of Ron and Mick. I said they were absolutely right to ascribe a politics to dead and stranded cetaceans, and maybe I needed to borrow some of their rightist confidence; not be so dismissive of my own experiences and convictions. How easy to dismiss and pathologise the personal symptoms of the mind – figure them as the discrete problem of the individuated, 'stranded' subject rather than symptomatic of something bigger; erase any potentially orientating circumference by always placing the emphasis on a malfunctioning self-centre. It's always the individual who's broken, the odd one out. Mark Fisher, the recently deceased cultural critic and author of *Capitalist Realism* (2009), argued that mental illness was not an individual issue to be treated and 'dosed', but indicative of a broader systemic sickness (bizarre that this even needs saying in our culture). Had I been the only one; had I lived among the dead whales in isolation, then maybe 'paranoid self-disintegration' might have served as a sufficient diagnosis. Maybe I would have approached my GP for a cure. But I wasn't alone. I'd been uniquely positioned as a ceti-receptor. The stranded whales were saying something about our world and it was my responsibility to listen.

The whales weren't valorising British nationalism – Brexit was just a highly successful and opportunistic ideological spectre-scavenger preying on large swathes of a stranded population. No, if anything these creatures were the symbolic embodiment of the post-2008 political and economic ebb tide that had prompted this resurgence of nationalism. *Of course* I suddenly couldn't see for dead and stranded whales: stranded individuals were everywhere – not only paying the price for the world-destroying mistakes of those working in high finance, but also being personally blamed for not being able to re-float themselves in the aftermath.

And now we had a government peddling hard-right neoliberal doctrines of competitive self-interest and the need to stem the flow of public investment, rather than curtail the activities of the supremely wealthy supposedly 'trickling down' their wealth. Every new cetacean arrival was a vindication – and materialisation – of the unfolding human catastrophe.

We'd retreated to a pub and I'd already finished my drink. Sandy ordered another round; asked if I was going to continue my quest. I told her about the message Mick and Ron had sent on my behalf. I didn't expect anything back from Blue – hoped I wouldn't hear anything back. I didn't want to end up like him. She said that wasn't going to happen – I didn't have the funds. 'See where the whales take you.' I said that, for the first time in a while, I felt better. Which was just as well because in other news, Alicia was expecting a baby. I assumed mine, but to be honest I wouldn't put anything past the whales. Sandy gave me her congrats, insisted on buying us a glass of prosecco each; said it was unlikely my wife was carrying the whale messiah.

5

Dawlish

Some three months later and I was beginning to feel confident. This wasn't about getting better – the language of convalescence no longer applied to my situation: it was about coming to terms; listening to the dead whales as they increasingly flexed their presence in the world, as they became more eloquent in their commentary. I saw history wheeling into the present – the stranding of an entire nation on the beach of imperial nostalgia. I assembled the evidence board, began piecing together a detailed picture of how the dead whales had been involved at every stage of the crisis, quietly informing, guiding, its conceptual parameters. My mind turned to the London whale – the one that swam up the Thames in 2006. I was there, lining the riverbank that day with the rest. At the time, most of us assumed it had come to bring us more news of the unfolding environmental collapse. That was true enough. But who could have known that this whale was also about to articulate the dimensions of the coming, intimately related economic catastrophe; who could have predicted its symbolic migration into the ethereal world of finance capitalism?

On Thursday, 19 January 2006, a juvenile northern bottlenose whale breached the Thames Barrier and began swimming up-river towards London. By the following morning, it had swum right past the financial district at Canary Wharf and as far as the Houses of Parliament. The first sighting of the whale in Central London was by a commuter who telephoned British Divers Marine Life Rescue. Crowds soon flocked to bridges and balconies, huddled at office block windows to catch a glimpse. Inner-city teachers

took their classes on impromptu whale-watching field trips. The whale became a celebrity.

> 'Whaley' – *Mirror;* FREE WILLY!! – *Daily Mail;* 'We shall call him Billy' – *The Times*; 'Wally' – *Sun* and *Daily Star*; 'Pete the Pilot' – *Evening Standard* (based on a misreport that it was a pilot whale); CELEBRITY BIG BLUBBER – *Sun*.

> WHALE STRANDED IN THAMES. A number of members of the public jumped into the river and splashed around to encourage the whale to move back into deeper waters... It appeared to have received a small cut from rocks, and earlier today there were signs of blood in the water.
>
> Mary Oliver, *Guardian*, Friday, 20 January 2006

It quickly becomes apparent that the Sunday souvenir editions planned for 'London's Whale' are going to be obituaries. A post-mortem reveals she –Billy, Willy, Wally, Pete, etc. – was female. A juvenile female. Six years old. There were multiple causes of death: severe dehydration, muscle damage from stranding-related injuries, and organ failure expedited by being hoisted in an inflatable raft onto a barge where she couldn't survive the pressure of her own weight. Starving. Her last meal: riverbed algae, a potato and a rubber glove from one of the would-be rescuers.

She is a Royal Fish. The *Sun* teams up with Her Majesty's Receiver of Wrecks and the Natural History Museum to raise money to preserve her bones for posterity. She becomes 'SW2006/40'. The SW stands for 'Stranded Whale'. The fortieth whale to come into the collection in 2006.

The Thames Whale: a celebrity specimen:

> There is little doubt in my mind that people have invested varying degrees of emotion in the Thames whale and what it represents to them. That ranges from it being emblematic of the state of the world's oceans, a struggle to survive by a creature trapped in the wrong environment or just a reminder that the wild is not so far from our doorsteps.
>
> Richard Sabin, Principal Curator,
> Natural History Museum, 2017

Agreed, Richard: the wild is not so far from our doorsteps. In vaporised form, it flows as capital through us all. Beautiful, socially complex creatures turned exchangeable commodities migrating through the air – as capital. Because under the capitalist regime 'all that is solid melts into air.' Anything located, rooted, sacred is ripped up, killed, expropriated – sold. It becomes part of a spirit world of exchangeable vanquished ghosts. The capitalised blubber of a billion dematerialised whales, combining with the blood of a billion stolen human lives, circulating through and defining all things. Nature is not far from our doorsteps.

We still live in the long shadow of the 2008 global financial crisis; the moment the transnational banking sector became socialised; when the most at-risk banks were rescued by the most at-risk people. Coming to power in 2010, the UK coalition government began slashing public funding and stripping away welfare and work benefits. The economic tide had turned. Financial liquidity had dried up. We entered the Age of Austerity – paying to sustain the force that was destroying the planet. The massive, upward redistribution of wealth. Bankers carried on as before.

As the first deep cuts were rolled out, the Thames Whale made a symbolic migration – into a Canary Wharf skyscraper. After amassing a $6.2 billion (£4.3 billion) trading loss for JPMorgan Chase, Bruno Iksil became known as 'the London Whale'. He is first on the list of all-time losers. Second is Howie Hubler of Morgan Stanley, and third Jérôme Kerviel of Société Générale. (Kerviel is the only one to have been convicted and imprisoned for their behaviour.)

What makes the London Whale stand out, though, is that his losses were incurred at JPMorgan Chase in 2012, four years after the global financial disaster. Naïve to assume that 2008 was a chastening moment for the banking sector, that oversight and regulation could curb the addiction to risk.

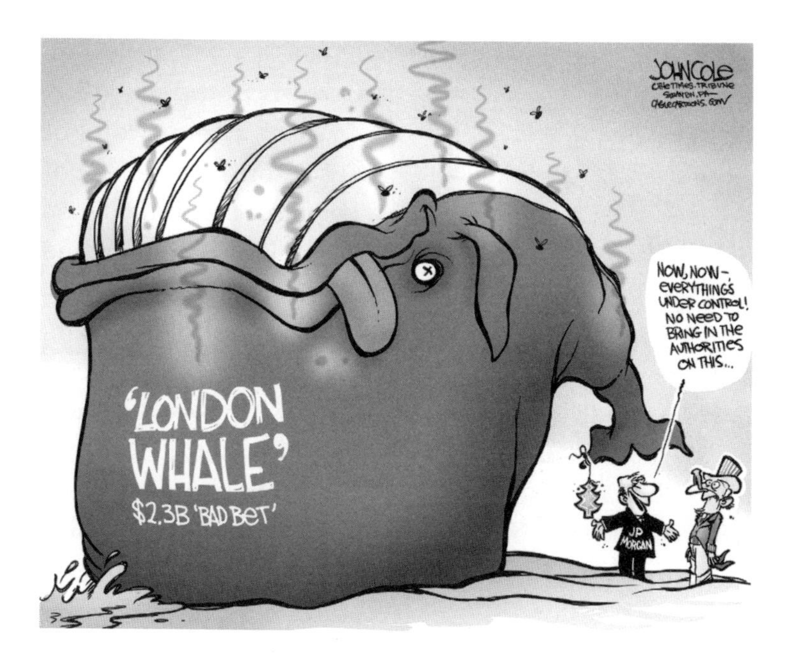

It is our duty to the American public to remind the financial industry that high-stakes gambling with federally insured deposits will not be tolerated. In 2012, the 'London Whale' trades resulted in a $6 billion loss. What if it was $60 billion? Or $100 billion? Does JPMorgan operate under the assumption that the taxpayer will bail them out again?

Senator John McCain, speaking at the Hearing before the Permanent Subcommittee on Investigations of the Committee on Homeland Security and Governmental Affairs United States Senate One Hundred Thirteenth Congress, 15 March 2013

"TIGHTER REGULATIONS?! YOU CANNOT BE SERIOUS!"

The whale trades demonstrate how credit derivatives, when purchased in massive quantities with complex components, can become a runaway train barrelling through every risk limit ... Firing a few traders and their bosses will not be enough to staunch Wall Street's insatiable appetite for risky derivative bets or stop the excesses ... The whale trades expose problems that reach far beyond one London trading desk or one Wall Street office tower. The American people [and the rest, mate] have already suffered one devastating economic assault rooted largely in Wall Street excess. They cannot afford another. When Wall Street plays with fire, American families get burned.

Senator Carl Levin, who oversaw the investigation

MONSTERS LIKE THE LONDON WHALE
SWIM IN WATERS TOO DEEP TO FATHOM.
Cartoon by David Simonds, 15 June 2012.

Not fair to pin it all on the London Whale. McCain continues:

> After these losses were uncovered by the press, JPMorgan chose to conceal its errors and, in doing so, top officials at the bank misinformed investors, regulators, and the public … During the same earnings call, Mr Dimon [Jamie Dimon, Chairman and CEO of JPMorgan] tried to downplay the significance of the losses by infamously characterising them as 'a complete tempest in a teapot' … The size of the potential losses and the accompanying deception echo the misguided and dishonest actions

> that the banks took during the financial crisis four years ago.

That's Jamie Dimon under the whale, holding a piece of paper that says 'Quarterly Results'. The woman in red is Ina Drew. This is her testimony at the Senate Hearing:

> This was my life's work. Through at least seven mergers and many financial crises, I always tried to do my best and what was right for the firm in a thoughtful, diligent manner. I loved the work ... On Friday night, May 11, 2012, I walked into the office of Mr Dimon, with whom I had a close and respectful relationship. I told him of my decision to resign from JPMorgan. It was a devastating and very difficult decision for me. It marked the end of three decades of hard work at an institution I loved.

Loved.

A 2017 *Financial News* article, 'The London Whale Resurfaces: Iksil Speaks Out,' interviews Bruno – now a stay-at-home dad stranded in the Paris suburbs. He is concerned that the epithet 'London Whale' has 'turned him into a scapegoat for a much broader problem'.

> BRUNO IKSIL: 'I WAS ONLY FOLLOWING ORDERS,' SAYS LONDON WHALE. The former JPMorgan Chase trader known as the 'London Whale' has broken cover to say he was not responsible for the scandal that lost the bank $6.2bn. In a letter sent late on Monday night to news outlets including *Financial News* and Bloomberg, Bruno Iksil said he was 'instructed

> repeatedly' by his superiors to carry out the trading strategy that led to the losses.
>
> Angela Jameson, *Independent*, 24 February 2016

Dimon tried owning the metaphor, taking inspiration from the political cartoons lampooning him in the aftermath of the scandal:

> JPMORGAN CHIEF JAMIE DIMON DEFENDS BANKS. 'Sure, we make mistakes, like we have got this Whale thing. Businesses make mistakes. So we've got to clean them up, learn from them, and get better. And I want you to know the London Whale issue is dead. The Whale has been harpooned. Desiccated. Cremated. I am going to bury its ashes all over.'
>
> Andrew Trotman, *Telegraph*, 13 August 2012

JPMorgan Chase continues to be a global leader in financing the climate apocalypse. Since the signing of the Paris Accords in 2016, and as of 2020, the bank has invested $316.74 billion into fossil fuels – making it the world's worst 'fossil bank'. (Source: 'Banking on Climate Chaos': Fossil Fuel Finance Report, March 2021.)

I want you to know the London Whale issue is dead. The Whale has been harpooned. Desiccated. Cremated. I am going to bury its ashes all over. A whale's appearance on a beach, a bewildering amalgamation of nature and economy: so irretrievably entangled it becomes impossible to prise apart one seemingly oppositional idea from the other. The stranded whale as the malfunctioning banking sector. Dimon rattled and rightly so – the whales of London provided yet another glimpse of the catastrophic toll. No wonder he was keen to erase the evidence. Desiccate, cremate. Bury the ashes all over. Vaporise. Dematerialise. Spirit away.

Stranded whales are an affront to the smooth functioning of capital and the amniotic bliss provided by the sense of liquidity that comes with knowing the whales are still swimming. A population rocked to sleep by the endless repackaging of the nature fetish and the pristinist delusions of the genre. To those who feel moved to come to the rescue: refuse the anaesthetising pleasures offered by the still-swimming whale and pick up an axe.

And they kept coming ashore, each one a further rhyming vindication of the strengthening hypothesis: the stranded whales speaking up to provide timely, trenchant critique.

Dawlish, Devon. Autumn 2016. Clive looks like an addict. He's not. He works for Deliveroo, biking back and forth to Exeter most days to deliver food to wealthy students. His face is gaunt and his teeth are mostly gone. I met him on a beach in Dorset a few years back. We were both staring at the newly arrived carcass of a bottle-nosed whale when his dog Costello, a Staffordshire bull terrier mutt afflicted with a number of skin diseases, came over to say hello; tried to take a salutary piss on my leg. Clive ran over to apologise. Within three minutes, he'd told me he'd boxed for the Army, had grown up in the grounds of a psychiatric hospital on the north coast of Cornwall, and that he was 'fucking poor'. A combination of warmth and nerves, he also began telling me about the various whales he'd seen come and go along this stretch of coastline; priding himself on 'making the effort for them' when they turned up on the South Coast, even if that meant biking miles.

I texted him late in September, asking if he'd already been to visit the badly decomposed fin whale that had just arrived. He suggested I bring beer; he'd collect some driftwood. We'd make an evening of it.

It'd died off the coast of France and had been slowly making its way across the channel, followed by a pack of sharks. The corpse was all over the local news: Whale at Dawlish. Smell awful. In terms of proximity, this is the closest a whale had ever got to me. It didn't come as a particular surprise – a whale so exactly zeroing in on my location and so very dead, too. It merely reconfirmed the weird precision of the forces at work here. It's rare for whales to get lost in the waters either side of the Exe estuary. A very young Edward stranded on Exmouth beach in 2002 – was forcibly refloated by holidaymakers – then stranded again on the other side of the estuary at Dawlish Warren a day later. But that's the only instance of cetacean miscalculation I could find. In the case of this Dawlish drift whale, it's as though the oceanic currents had conspired with the spirit whales to place this corpse as close as it might reasonably come to my location. Only polite that I greet it.

With enough alcohol in tow for this impromptu wake, I boarded the train at Exeter St Thomas for one of my favourite rides in England. The track follows the right bank of the Exe estuary to Dawlish Warren and the sea, stopping only at Starcross, and then on to Dawlish Warren, Dawlish, Teignmouth, Newton Abbott, Torquay (fictional home of *Fawlty Towers*), and Paignton. The whole journey takes about fifteen minutes, much faster than you'd ever be able to drive it. An improbably beautiful escape from city to sea. Backlit by the afternoon and evening sun I counted oystercatchers, curlews, godwits, dunlins and Brent geese picking their way through tidal mudflats. Local fisherman waded the tideline among the ancient ribs of stranded fishing boats checking their traps for crabs, flounder and ragworms. On the other side the train trundles past the 600-year-old Powderham Castle, besieged by Cromwell's troops during the English Civil War, but no one ever remembers to look out for it.

I got off at Dawlish Warren, an enormous sand-spit nature reserve jutting out into the Exe estuary. All these places are wonderful, particularly at low tide when endless stretches of burnished Triassic sand allow you to walk out for miles among the clam and mussel beds. The Warren points towards Exmouth, and there's a links golf course cut into the lee side of the dunes where a Second World War machine-gun turret points at the first tee. You can still get a crab sandwich for £3.50, and if you look past the brownish iodine solution that's often used to clean the meat, it's definitely worth that. Much of this area is shielded by a persistent class snobbery and rumours of used heroine needles littering the beaches. Nearly everyone agrees: that's absolutely fine.

As the train doors opened, the familiar stench of death. Even at this distance it was like standing next to a distillation of the ripest fish-tripe summer rot. Incredible. It'd been dead for a long time – more than a month, probably. Children complained and held their noses on the station platform; adults bowed their heads as though walking against a stiff breeze. I passed an almost deserted funfair and penny arcade on my way to the coastal path, the ebb tide beginning to reveal the dunes that sweep in one continuous band across the bay. Kitesurfers and windsurfers in the distance at Exmouth, their sails a string of bunting against the dried-blood-coloured cliffs. The path here follows the railway line along the water's edge, punctuated by public information signs on the legacy of Isambard Kingdom Brunel – 'Visionary British Industrialist' – tagged in pink graffiti.

Three teenage boys sat outside an abandoned café. I must have been grimacing because one of them told me it was 'down there'. In the middle distance, twenty or so people and their dogs stood upwind of a badly discoloured corpse. About the size and shape of a partially melted minibus. A crime-scene fence had already been set up, though the encroaching flood tide was forcing the fluorescent-jacketed attendant to continually revise its perimeter. A group of kids inevitably daring each other to get as close as possible, ducking behind the body for cover. Two trains, one on its way to Paignton, the other to London Paddington, crossed at the whale – shielding the capital city commuters' view of the carcass.

Whenever I hear that sharks have been involved, my brain always splices two pieces of writing together: the shark massacre chapter of *Moby-Dick* – the crew of the *Pequod* in their beds hearing the snapping jaws of hundreds of sharks devouring the dead whale inches from their sleeping heads; and the blue marlin skeleton that Hemingway's Old Man finally brings back to shore. This body looked fairly untouched to me. I picked out Costello, trying to work out what part of this decomposing animal he might reasonably lick. Clive wasn't far behind, shouting at him to come away,

or at least move upwind. He picked up Costello by the scruff of his neck and gave him some stern words – before cradling him like a baby and giving him a scratch. The body a glowing luminous-rainbow putrescence in the evening sun.

Clive caught sight of me at a hundred metres or so but held back on the leeward side of the body. He was striking a pose and pantomiming wafting his hand in front of his nose. Behind the whale, the coastline curved towards Torquay – Hope's Nose, Babbacombe beach – the continuity of the land eventually punctuated by the full stop of Oreston Rock, where Henry James once sunbathed, and Oscar Wilde holidayed with Bosie. Fashionable resorts.

'Can't tell if it's Costello or the whale,' was Clive's greeting. I asked how they were both doing. He didn't think air molecules could move upwind, but it'd probably stink wherever we sat; pointed at a carefully prepared pile of driftwood he'd collected. About thirty metres away from the body, his bike and delivery box propped up against the sea wall. I asked if we should move a bit further back and he fished some VapoRub out of his pocket – 'a bit *Silence of the Lambs*, but it works'. A classic olfactory combination, at least among our people: eucalyptus and whale-corpse rot.

He'd brought along some 'specialists'. A really solid crop of mushrooms on Dartmoor last year. Not too potent. I peered down at the stash: enough to be interesting. The sun well down behind the cliffs now. We sat on the sand, leaned against the sea wall, lit a modest (semi-legal) driftwood fire and Costello turned in for the night. Save for one or two curious dog walkers, we were abandoned to our whale. Clive already had his phone out: a four-pound bull huss shark he'd caught last year and eaten ('two pound of meat off of that'). Then a picture of his mate's new bike. And then segueing, inevitably, to the whale-corpse vids. An old favourite: the episode of *Family Guy* where Peter Griffin arrives at the scene of a stranding driving a forklift truck, and announces to the assembled crowd that he's there to save the whale. Peter has trouble manoeuvring the vehicle; proceeds to completely ruin the carcass. Cue deadpan crowd scene where everyone stands in shocked silence. Peter announces, triumphantly: 'You're whale-come.' Then the ritual deep YouTube dive: exploding whale compilations. One clip: an entire city street – cars, windows, children on their way to school – splattered in blood and viscera. In 2004, a whale exploded in the Taiwanese city of Tainan: the decomposing body was en route to a nearby research centre. Methane gas had accumulated in the whale's stomach and the carcass's weight had closed off all available escape routes. The blast radius was 50 metres. Then finally (always) the 'Exploding Whale', one of the earliest viral videos (eventually featured on a *Simpsons* episode). November 1970, local authorities packed 20 cases of TNT under the whale's body in the hope of blowing it back out to sea.

The twelfth of November, 1970: a report by Paul Linnman of KATU (Portland, Oregon). Film by Doug Brazil:

> *[Shot of anchorman Linnman. Whale in background, slumped head positioned near the microphone, as though it were about to offer an interview.]*
>
> It had to be said, the Oregon State Highway Division not only had a whale of a problem on its hands, it had a stinking whale of a problem. What to do with one forty-five-foot, eight-ton whale, dead on arrival, on a beach near Florence.
>
> *[Cut to landscape shot of whale.]*
>
> It had been so long since a whale had washed up in Lane County, nobody could remember how to get rid of one.
>
> *[Cut to slow panning shot of the whale's body, starting with the tail.]*
>
> In selecting its battle plan, the highway division decided the carcass couldn't be buried because it might soon be uncovered. It couldn't be cut up and then buried because nobody wanted to cut it up.

[Cut to man in orange fluorescent jacket laying dynamite charge.]

And it couldn't be burned. So, dynamite it was – some twenty cases, or a half-ton of it.

[Cut to close-up shot of boxes of dynamite, and then further laying of charge.]

The hope was that the long-dead Pacific grey whale *[it was a sperm whale]* would be almost disintegrated by the blast …

[Cut to more prep.]

… and that any small pieces still around after the explosion would be taken care of by seagulls and other scavengers.

[Cut to still shot of whale in middle distance, and then a slow close-up pan to seagulls circling above it.]

Indeed, the seagulls had been standing nearby all day. As everything was being made ready, we asked George Thornton, the highway engineer in charge of the project, for his final observations.

[Cut to interview with George Thornton, wearing a white hard hat, orange jacket and light blue T-shirt. As he speaks a child kicks sand at him in the background, as if in warning.]

[George Thornton] Well, I'm confident that it'll work. The only thing is we're not sure just exactly how much, uh, explosives it'll take to disintegrate this … thing, so the scavengers, seagulls and crabs and whatnot can clean it up.

[Paul Linnman] Is there any chance it might be more than a one-day job?

[George Thornton] Uh, if there's any large chunks

left, and, uh, we may have to do some other clean-up, possibly set another charge.

[Cut to footage of boxes of dynamite being positioned under the whale.]

[Paul Linnman] The dynamite was buried primarily on the leeward side of the big mammal so as most of the remains would be blown toward the sea.

[Cut to footage of an older couple sitting under a grassy sand dune being approached by an official.]

About seventy-five bystanders, most of them residents who had first found the whale to be an object of curiosity before they tired of its smell, were moved back a quarter of a mile away.

[Cut to three more people standing in the dunes.]

The sand dunes there were covered with spectators and landlubber newsmen …

[Cut to more on the beach.]

… shortly to become land-blubber newsmen. For the blast blasted blubber beyond all believable bounds.

[Cut to shot of beach, just prior to explosion. Ten-second countdown. Huge explosion the colour of gore. A mother, off-camera, can be heard saying, 'All right, Fred, you can take your hands out of your ears now.' And then you hear the crowd's response turn from excitement to fear as whale meat starts raining down. Another woman says, 'Here comes pieces of, uh, whale.']

Our cameras stopped rolling immediately after the blast. The humour of the entire situation suddenly gave way to a run for survival as huge chunks of whale blubber fell everywhere.

[Cut to beach – a cloud of whale still visible.]

Pieces of meat passed high over our heads, while others were falling at our feet. The dunes were rapidly evacuated as spectators escaped …

[Cut to parked trucks.]

… both the falling debris and the overwhelming smell. A parked car over a quarter of a mile from the blast site was the target of one large chunk …

[Cut to smashed car, passenger seat completely caved in, with the trunk sticking up.]

… the passenger compartment literally smashed.

[Cut to another shot of the damage, the chunk of whale meat bright red, and surprisingly small, indicating its velocity.]

Fortunately, no human was hit as badly as the car. However, everyone on the scene was covered with small particles of dead whale.

[Cut to shot of man taking a picture of damaged car.]

As for the success of the effort, well, the seagulls who were supposed to clean things up were nowhere in sight, either scared away by the explosion or kept away by the smell. That didn't really matter.

[Cut to shot of bulldozer burying the rest of the carcass.]

The remaining chunks were of such a size that no respectable seagull would attempt to tackle them anyway.

[Cut to Linnman and his crew talking to George Thornton, who is now looking at the floor. Back to bulldozer rolling a chunk of the corpse forward.]

As darkness began to set in, the highway crews were

back on the beach burying the remains, including a large piece of the carcass which never left the blast site. It might be concluded that, should a whale ever wash ashore in Lane County again, those in charge will not only remember what to do, they'll certainly remember what not to do.

[End scene.]

The blast blasted blubber beyond all believable bounds. An amazing line. See the precise moment when the immovable monumental weight of a cetacean corpse bursts into air particles – a gigantic cloud of whale rising up above the beach, only to apparently condense in a hailstorm of flesh. A whale storm. A weather system. A vaporised cetacean phantom recombining in mid-air. A three-minute replay of the last four hundred years of capitalocene culture: the upending of all believable bounds – the subsequent disorientating recombinations of matter. The certainty of a whale's body and suddenly parts of it are landing on cars – spectators fleeing from the whale droplets raining down from the sky. The material reality we've come to collectively inhabit. The corresponding recalibration of the kinds of sentences and thoughts this culture was forced to think and articulate: the blast blasted blubber beyond all believable bounds. The stranded whale as ground zero, the site of the detonation.

Clive came back with more fuel, the tension relaxed a little from his expression. He looked thoughtfully at his bike, drew in a lungful of sea air. With Dartmoor specialists, the fear and paranoia nearly always come (usually the next day), but you're also handed a few moments of presence, a period of grace that justifies the repayment plan. The gentle pulsing light on the waves, a dead whale's corona. Clive began by telling me about the tip he'd received that day. A tenner. One of the biggest he'd ever got. He paused, riding a slight psychoactive transition. He said he didn't like school – wasn't stupid or anything – just didn't come online early enough; got interested in things when it was a bit too late. He was fifty next year. I hid my surprise in a gulped swig of beer. Late thirties, I'd thought.

'I know how to ride a bike and scaffold.' A couple of years ago, he dropped a metal pole on someone's head and that ended the scaffolding. I asked him how long he'd served in the forces. Four years. He asked me if he'd told me about the fight in a Hong Kong bar during one of his tours. I said he hadn't, but he'd told me the story every time we'd met. The scar was on his right arm; he'd blocked a chair leg and landed a punch with his left. Got out of there quick.

Clive stood up and I joined him. A round of the whale's perimeter. I'm not sure I can think of another member of the subculture who is so committed (and eloquent) in their self-abasing identification with them. For him, this was always a meeting of outcast kindred spirits. Standing next to an old dead whale, the scars across their bodies speaking to the scrapes and adventures of a lifetime, Clive said he felt more sympathy with these creatures than with any human. He had no qualms about standing next to a whale and reminiscing about the good old days, lamenting the situation they both currently found themselves in. Clive would stop, put a hand on a rotting fin, and directly address the corpse. *For Orpheus' lute was strung with poets' sinews,/ Whose golden touch could soften steel and stones,/ Make tigers tame and huge leviathans/ Forsake unsounded deeps* *to dance on sands.** The night extended itself; trains shot past our heads; the specialists had come to a polite stop at the edge of something more serious, the hard edges of the visible world reconstituting themselves gently in the psychonautic breeze. The whale's taut skin an astral mirror.

Clive said I looked uneasy. I started talking about my recent experiences, addressing both of them – whale and man. Asked the whale directly by what method she'd tracked me down. Save from swimming up the estuary itself, you couldn't have arrived at a more convenient location for the visit. Why did you want to be near me? What were you trying to say? Initially I'd just assumed it was a breakdown. More recently (she was the latest arrival to confirm the hypothesis), I seemed to be living in a world in which the iconography of a dead whale was breaching my every waking and sleeping moment – actually determining my thoughts, traversing and connecting up psychology, politics, economics. I asked them if they'd heard of Mick and Ron. Clive hadn't; the whale remained silent. Two blokes who live along the coast here, absolutely convinced the whales who turned up at the beginning of the year were trying to tell us to vote Leave. I gave my alternative read, my emerging spiel on strandedness, on austerity; felt confident enough to share my thoughts on the gig economy; how those who ride bikes delivering food to the rich do so without any protections or safeguards; how neoliberal capitalism specialises in making everyone 'self-employed' – subcontracting care itself, refiguring it as 'self-care', the individual's responsibility. No sick pay, no unions: the age of austerity, utterly exposed, stranded. Addressing the whale,

* William Shakespeare, *The Two Gentlemen of Verona*, III.2 (1623).

I offered thanks for providing such a glaringly obvious manifestation of our own society's economic insistence on periodically running aground. No wonder certain members of the community wanted to snatch the occasional body part: they were trying to hold open the momentary rupture in the fiction of economic prosperity and 'development'. The whales who still swim in the ocean sing too sweetly. Dead whales tell it true; the more mutilated the carcass, the better. Listen to the whales.

'*Da*, comrade', said Clive, playfully. Sorry to say, but riding deliveries was the only viable option he had to survive – in some ways couldn't help but feel grateful for the 'chum'. Not unkindly, he pointed out that we'd each had *very* different experiences of the last ten years. I stopped talking for a while.

I found that my head was changing gear, accelerating in a different direction; asked Clive if he'd ever taken part in a 'whale removal' – said I could be quiet if needs be. He was good about it, said the mushrooms had been his idea, and waved me on.

They come like thieves in the night to rid our collective

consciousness of the evidence; maintain the fetish character of the English Riviera. Had I told him this story before? I don't think I had (I had). Not properly, though. At the beginning of this decade I'd taken part in a whale clean-up. I was getting over my first significant relationship. It ended up with me wearing a gas mask, helping fasten the body of a whale into a giant harness, before watching a crane winch it into a lorry for disposal at a local landfill site. An experience unwittingly paid for by my Arts and Humanities Research Council PhD funding. Three years of twelve and a half thousand a year, paid quarterly. Money that meant I didn't have to make any immediate decisions about my future, and allowed me to seamlessly float from one degree to the next – read more Herman Melville and his kin.

Clive punctuated this with, 'Lucky fucker'.

I agreed.

We met each other at a loud house party and woke up together the following morning. I took her to a seafood restaurant, ordered a platter, fell hopelessly in love. From her bedroom window I could see the light flashing on top of Canary Wharf and I pictured the liquid Thames flowing underneath – the precise spot where 'London's Wonder' had come to die. I was due to go on a research trip, to New Bedford, Massachusetts. I'd been awarded the Herman Melville Archive Fellowship – I'd get to take part in the annual marathon reading of *Moby-Dick* – and stay in the museum's guest rooms.* I returned to England, handed over

* The trip led to my first academic publication. See Peter Riley, 'Report: Melville Society Archive Fellowship', *Leviathan: A Journal of Melville Studies*, 12.2 (2010), pp. 107–10.

the matching T-shirts I'd bought, white whales swimming across the front of them. The relationship lasted another six months. It ended on a Sunday evening; she said I looked sad. I was the opposite of sad and asked if she wanted to end it. She nodded. I got on a night bus, changed at Piccadilly Circus, caught another bus to Kingston-upon-Thames – no more buses – and walked the three hours back up along the Thames to my mother's house in Staines.

'Back to mum', Clive said, picking up Costello and putting him on his lap for a scratch. The dog sighed with pleasure.

'She wasn't in,' I lied.

'I bet she was,' he said. 'I bet she gave you a hug and cooked you breakfast. What mums are supposed to do.' I didn't say anything. He raised a bottle to the stars. 'Here's to mums.' We both swigged. He stopped for a train to pass, told me to carry on. He was having a nice time, indicating his chest with a cocked wrist and extended fingers. 'You were heartbroken.'

The months passed. I stopped doing my work, narrowly passed my annual review – was going to drop out. I waited and waited for the next arrival, certain that a dead whale ought to form part of the grieving process. Nothing but a couple of minor incidents up in Scotland. At last, a sperm whale: Thanet. I'd creep up on it, walk the coastal path north from Dover, do some camping and processing. I didn't really know that part of Kent – Deal, Ramsgate, Margate, Whitstable, those kinds of places. I checked and packed my old camping gear – a discreet and waterproof one-man tent, a sleeping bag, stove, and my lightweight mackerel rod for good measure – and caught the afternoon train to Dover. It was raining when I arrived and I didn't stop walking until I got to Deal some ten miles away. The sky had cleared with about an hour of daylight to spare. At the pier's end I sat down on a bench, opened a tin of beans and ate them out of the can. For dessert, line-caught tuna pieces in brine. I put the tins in a plastic bag that I tied to my rucksack. A man next to me pulled up a decent-sized, orange-burnished pouting, let it flap around on the floorboards as he baited his hook, eventually kicked it back into the sea. I watched it sail through the air and belly-flop into the water.

A turning tide, late summer warmth obscuring the French coast that ought to have been visible in the distance. I stationed myself at a location from which I could cast both towards the beach and out to sea. Decent lead sinkers for distance; a tangle of six barbed and rusted hooks with the feathers dyed green and yellow. Mackerel rod still caked in the fish scales of the previous excursion. To my left, a teenage girl was already lifting three large fish over the handrails. She transferred them into a bucket, unhooked them quickly. The bucket rattled intermittently along the boardwalk and came to a full stop.

Patience and about fifteen minutes before I was untangled. First cast, nothing. Second cast and the rod-tip bowed. Healthy-sized fish, their combined panic cancelling itself into a dumb weight that I started mechanically reeling in. A sudden gust of wind unfastened the rubbish bag from my rucksack and blew it into the sea. I watched it parachute among the waves, apologised, and felt like shit.

Four beautiful mackerel, the last rays of the sun setting their oil-slick blue-black skins aflame. Hunting eyes crystal-clear. I grabbed one and felt the shock of its electricity and strength. I held on tight: the hooks were too deep. When I'd finished with them, they lay there in ruin. I hid the evidence in another plastic bag, picked up the rod and made to cast again. The bail arm wasn't set. As I flung the weight out, the line audibly snapped and my rig sailed up into the sky and then dropped into the sea. The man who was fishing next to me pointedly fixed his stare onto the horizon. I gathered up my stuff along with my four little corpses. The girl, no older than fourteen, came over and offered to share some of her catch. 'I'll get plenty more,' she said, about fifteen already in her bucket. Without thinking, my plastic bag opened, and she doubled my haul. She asked me how old I was. Twenty-four. She frowned and moved back to her spot.

Clive had closed his eyes. Fair enough. I asked if he was still listening. He hummed assent and waved me on. Costello sighed deeply, rearranged his tongue. Clive, the bike, the delivery box, the dog, the fire, the sea, the stars, the whale. And then among the constellations, three other forms sitting together, listening in. Not complete materialisations by any means, but tangible presences nonetheless: combining among the shadows cast by the embers of the fire Theo, Nicholas Redman and the blue-haired comet woman. I addressed them all.

The eight mackerel kept me company for about half a day, and then I chucked them under a bush; hoped they'd be a treat for a fox. After a few paces, I doubled back on myself and covered the fish with some leaves – an attempted burial. A further day slipped by without notice; an exhausted night sleeping on a park bench. It occurred to me that I should see this one through from start to finish – watch, maybe participate in, the process of the whale clean-up itself. Go through the motions. A fellow walker, coming in the opposite direction, confirmed the news; told me about the sperm whale stranded a few miles along the coast. Had been there for a few days now. Did he know when they might take it away? Not a clue.

As with the Norfolk coastline, the stretch between Dover and the Isle of Sheppey to the south-east of London is another of those places where whales have traditionally come to die. The Swale, the treacherous moat that protects Sheppey, has an extended history in this regard, being yet another of the great stretches of tidal flat in Britain. Richborough Roman fort – another few hours walking by my calculation. As late as the Middle Ages, the district of Thanet was actually the Isle of Thanet.

'Nigel Farage, leader of the UK Independence Party, contested South Thanet in the 2015 General Election and won 16,026 votes' – the Dawlish whale said this thoughtfully, and with her back to us so that no one could tell whether her jaws were moving as she spoke.

All foreign pilgrims – including whale pilgrims – were required to sail or swim around it if they wanted to worship at Canterbury. The channels eventually silted up, and the island became part of the mainland. In the Corona Chapel of the cathedral itself, there's a thirteenth-century stained-glass representation of Jonah being vomited onto land. Surely the local monks commissioned it because of the whales that inevitably and persistently tried to swim up the River Stour towards them (at least as frequently as they swam up the Thames). There is a record of a pilgrim sperm whale beaching on the Isle of Thanet in 1574:

> The IX of July at six of the clocke at night, in the Ile of Thanet besides Ramesgate, in the Parish of Saint Peter under the Cliffe, a monstrous fish or Whale of the Sea did shoote himselfe on shore, where for want of water, beating himselfe on the sandes, hee dyed about sixe of the clocke on the next morning, before which tyme he roared, and was heard more than a myle on the lande.

The length of this whale was twenty-two yards, the 'nether iaw' twelve. Local scavengers seem to have been particularly preoccupied with the creature's giant eyes – easily big enough to spot a sinner when they saw one: 'One of his eyes being taken out of his head, was more than sixe horse in a cart could draw, a man stoode vpright in the place from whence the eye was taken.'* The fantasy removal of a whale's

* In Raphael Holinshed's *The Firste and Laste Volume of the*

gigantic eyeball, a gouging out, a climbing down into the eye socket, a passing through the windows of the non-human soul. Standing there, looking out. What does this creature know? 'Lest it see more, prevent it. Out, vile jelly!' (*King Lear*, Act III, Scene vii).

Another few hours of steady walking before I reached the carcass. Forty foot or so long, it had been washed up close to a dirt track. About fifteen people were gathered round, and I added myself to their number. A hi-vis-jacketed man holding a clipboard was explaining that the carcass would be removed to landfill tonight. I waited for a lull in the dialogue and approached Hi-vis. In his forties – glasses and a black beard streaked with silver and lunch. I said I wanted to help.

'Yeah, we all want to help, mate.' He wrote something in pencil on his clipboard.

No, not like that. I wanted to help with the removal. He looked at me suspiciously, asked me why I'd want to do something like that.

It was just something I'd always wanted to do.

He looked up over his glasses. 'You're a journalist?'

No, I just wanted to help with the actual work; moving it onto a truck and unloading it.

Chronicles of England, Scotlande, and Irelande Conteyning the Description and Chronicles of England, from the First Inhabiting unto the Conquest (London, Iohn Hunne, 1577). See also the 'Description of the Sperma-ceti Whale, thrown on the Flats at Sea Salter, near Whitstable in Kent, Dec. 1763' in David Henry (ed.), *The Gentleman's Magazine: and Historical Chronicle*, January 1764: 'The extreme length of the fish was 54 feet; its girth in the broadest part over, back and belly 38 feet ... This fish being thrown on the manor of the dean and chapter of Canterbury, was sold by them for 80l.' FYI: Lovely illustration of this whale easily locatable online.

'Sorry, mate, you'd need to sign up to an agency.'

'What would they do?'

They'd insure and train me up for a start. And I probably wouldn't be able to choose what I cleaned up.

They'd put you on the public shitters before they'd put you on one of us, mate. Redman and Blue Hair smiled broadly at the whale's comment. Even Clive, who was mostly asleep now, nodded his appreciation. I waited for them to settle down, continued my story.

I offered the guy fifty quid. He looked at me. Then the whale. After a moment's pause, he asked if I had the money on me. I could get it, provided there was a cash machine nearby. Ten minutes later, I'd loaded my bag into the trunk and was sitting in the passenger seat of his white, newly valeted Kia. A Cross of St George air freshener dangled from the rear-view mirror. On the drive he asked if it was 'something sexual?' I explained I'd just come out of a relationship, and that this was how I'd decided to process things: clear beach, clear mind. He made a noise of half-interest. I asked when he was coming back for it. It wouldn't be him tonight – he did the risk assessments. After a few moments of silence, he told me I'd probably have to make it a hundred. 'Have to make some more payments down the line' – make absolutely sure it was all above board. We parked outside a local convenience store. I saw the arc of what was coming minutes before it happened.

'Does this machine charge?' I said, more annoyed by this than the prospect of being mugged. He didn't know, but I should hurry up now if he was being honest. I took out a hundred – it would mean I'd need to make one or two cutbacks over the coming weeks; got back into the car; handed over the money. He took it, shoved my bag in my lap, opened the passenger door and told me to fuck off. I obliged. The window came down fractionally: 'Queer.' He drove away, wheels skidding slightly as he started.

Without pausing to think I went back to the cash machine, took out a further fifty – then asked in the convenience store where I could buy a hi-vis jacket and a hard hat. The teenager behind the counter said he had no idea. Eventually a bus pulled up on the other side of the road. When I told the driver where I needed to be, he said he was going in that direction, though it would still be a good walk; if I waited for the next one, it'd take me closer. I got on the bus and it dropped me off about two miles away. The walk was difficult, along a narrow road guarded by high hedgerows – shut in by the slate-grey English sky; forced to become part of the hedges when passed by cars. Lying on the seat of a parked-up, and unlocked, tractor, a hi-vis jacket. I stuffed it into my bag. If you want to be taken seriously near a whale, you need the hi-vis.

Almost sundown, and the whale was alone now, its jaw intact, body bulging in volatile rot. As I approached, the unmistakable song of the beached sperm whale, its death gurgle. Escaping gas from various compressed apertures. Jaw wide open as it would go. If they had any experience of the process, the removers would make an incision into the stomach and stand well back. In some instances, this could mean the difference between hoisting a large bus and a small car.

I put down my things, orbited; found the only place you might safely lean on a decaying sperm whale – the tip of the forehead, or 'battering ram', as Melville has it. Solid enough. I looked out across the water. An ebb tide the reason they were coming tonight. The smell was astonishing, dementing. I scanned along the length of the body, traced the scars of deep-sea battles with giant squid looped in white oceanic calligraphy along the length of its body. I watched it join up with human lettering. 'AJ 4 PF' encircled by a heart. Scraped into the flesh. Arborglyph, the writing you find on trees. This was cetiglyph. Not without precedent in Britain.*

The impulse to defacement, dismemberment. Revolting. Why deny the whale its natural stateliness, even in death?

Natural stateliness. How perversely comforting it is that there are still some whales left to strand – cetaceans enough to arrive and broadcast such steadfast completeness. Lying there whole. On the days I'm able to think past my gut reaction, I see the morbid integrity amidst the carnage – a making sure that the whale can no longer be mistaken as part of a pristine, natural world – an ocean blue that might yet be untangled. Whales' bodies – our bodies – always *already* a palimpsest articulating the combined, violated fates of different species. All of us variously tagged and dismembered, not by petty vandals or deep-sea squid, but by another monster, the stranded whale's symbolic and deadly counterpart.

---

* In 1943, a divisional court upheld an appeal by a couple who had been prosecuted for 'carv[ing] their initials in the *living flesh* of whales stranded by the tide on a shore in Devonshire'. The defence had successfully argued 'there had been no cruelty by reason of the fact that though the law protected domestic animals and wild animals in captivity, the whales did not come within either of those categories in that they were stranded simply on the shore by the forces of nature.' See 'CONSIDER ANIMALS IN THE POST-WAR SCHEMES,' *Surrey Mirror*, Friday, 23 April 1943.

> The first thing you need to know about Goldman Sachs is that it's everywhere. The world's most powerful investment bank is a great vampire squid wrapped around the face of humanity, relentlessly jamming its blood funnel into anything that smells like money. In fact, the history of the recent financial crisis, which doubles as a history of the rapid decline and fall of the suddenly swindled dry American empire, reads like a *Who's Who* of Goldman Sachs graduates.
>
> Matt Taibbi, 'The Great American Bubble Machine: From tech stocks to high gas prices, Goldman Sachs has engineered every major market manipulation since the Great Depression – and they're about to do it again', *Rolling Stone*, 5 April 2010

Damaging caveat: this political metaphor is entangled in the history of nineteenth- and twentieth-century anti-Semitism. See for example the anti-Semite, racist and white supremacist H. P. Lovecraft's writings on the cephalopodic monster

'Cthulhu'. There have been various attempts to rehabilitate/propose a new, less historically inconvenient version of the metaphor since 2008. In the wake of Taibbi's article, there was the 'Fight the Vampire Squid' slogan of Occupy Wall Street (Octopi Wall Street), and then more recently in 2019 Extinction Rebellion's massive pink inflatable octopus paraded down Whitehall.

> Just as the cuttlefish squirts its ink in order to protect itself, [capitalist ideology] cannot rest until it has obscured the ceaseless making of the world, fixated this world into an object which can be for ever possessed, catalogued its riches, embalmed it, and injected into reality some purifying essence which will stop its transformation, its flight towards other forms of existence … immobilise the world … suggest and mimic a universal order which has fixated once and for all the hierarchy of possessions.
>
> Roland Barthes, *Mythologies*, translated by Annette Lavers (1972)

Eventually they came. One o'clock in the morning. Five men. A lorry, digger, and small lifting crane. Jacketed, I strode up to the person who was directing the operation and stated that I was here to help, even tried apologising for being late. Before he could tell me to leave, I offered him the last fifty quid, which he took without hesitation. I'll call him Colin. He generously told his colleagues that I was on work experience. For the money, he gave me a gas mask, said to put it on when I was told and stand well back. This wasn't for the faint-hearted. He also informed me that I absolutely could not get injured. I wasn't insured, so anything worse than a scratch and he would haul me to the landfill site as well. 'We'd hide your body in the whale, do you understand?' He patted me on the back.

The youngest-looking member of the team handed me a halogen work light and told me to point it at the whale's stomach – at a 45-degree angle, about ten metres away, and upwind. The night was relatively still. Two of the men laid out harnesses so that they were perpendicular to the carcass's underbelly. 'Is that square?' one of them called out to me. I said it was. They were going to try and roll the body over and onto them, and then lift the whole thing into the truck. The digger came forward with its loader bucket and stationed itself at the mid-point of the whale's spine. Madness – the whale would explode.

The driver switched off the engine and got out. Colin walked over to where I'd positioned the light and told us to stay where we were. He was carrying a shotgun. He loaded both barrels; told us to put our masks on. Gas mask, double-barrelled shotgun, whale backlit by halogen. The gun kicked hard against its human brace, echoing a whip-crack of lead shot into blubber. We waited and listened. Nothing but the steady death gurgle. Colin aimed and fired again, hitting roughly the same spot. He took two paces forward, reloaded. Before he pulled the trigger a third time, the whale stirred. A great rush of boiling gas began releasing from the wound.

A slow-motion olfactory detonation. The equivalent of staring directly at the sun. I started retreating, ripping my mask off in the process. I didn't know I had the capacity to smell that much. My colleagues retreated too. I went down on my knees, retching like the child from *The Exorcist*, every sinew in my body straining to rid itself of this evil. The turd wranglers of London, daily wading through rivers of shit and blood in the Middle Ages, would surely have been chastened by this hell-mouth opening. Only Colin held his ground, but even he was down on his haunches, palm bracing his masked forehead, shotgun pointed to the sky (his is still the most elegant solution to whale disposal I've ever come across).

The carcass took about a minute and a half to stabilise; me a little longer. The diameter of the wound was maybe the width of my shoulders (honestly the width of my shoulders – that's not a retreat to Freud). The digger then started moving slowly forward. It hoisted its bucket and rolled the massive weight onto the harness bands. The crane came forward, and I saw the tiredness and concentration on the driver's face. The lorry moved into position and the crane did its work. The mass of burping, dripping flesh soon found its hearse.

'Mind how you go,' was Colin's cheery farewell.

Had it done the trick? asked the Dawlish whale. *Had the experience helped me process the break-up?*

I said I'd soon started feeling better – 'Make of that what you will.'

The whale didn't ask any further questions. My other listeners were congealed into one entity now, perched at the ledge of my peripheral vision, a shadow on the sand. I watched them slumber, the certainty of the carcass continually at play among the fractions of light.

I woke up. The fire was out, and the eastern clouds glowed morning. Clive and Costello, the rest of the gang – save the whale – were gone. My bag under my head for a pillow. I'd had an undisturbed night's sleep. A small group had already assembled – one or two threw concerned glances my way. At some point during the night the breeze had apparently changed direction. I smelled it now in a way I hadn't the night before. Overpowering, incredible. I got up, brushed the sand off and walked back to the station. I had seminars to teach on Monday, the term in full swing. Convinced that I was saturated by whale stink, I turned up ten minutes early to my first seminar and made sure that all the windows were opened as wide as they would go. I then clustered my students' desks as far away from me as possible and taught from the windowsill.

Introduction to Poetry – the week on popular balladry. Only fitting we talk about William McGonagall's 'The Tay Whale'. Had anyone visited the whale just washed up at Dawlish? Just me, then. You should take a look: it's probably going to be removed later on today.

Lesson plan/collection of sources for William McGonagall's 'The Tay Whale' – held up as one of the worst poems ever written in English:

Arrival of humpback whale in the River Tay – Dundee whaling fleet down on its luck and with point to prove – whale hunt as public spectacle – many heroic escapes – popularity and poetry – prolonged carnage – washes up dead – sympathy for the whale, rehabilitation of his 'Whaleship' – rehabilitation of McGonagall. On the intersections between economics, precarity, and dead whales.

Preamble.

FAILURE OF THE WHALE FISHING ... 55 whales, against 212 whales last year and 225 whales in 1881. The result of the whale fishing this year is therefore the poorest on record for both the Dundee and American fleet.

Dundee Advertiser, Tuesday, 6 November 1883

Part I: First sighting.

'Twas in the month of December, and in the year 1883,
That a monster whale came to Dundee,
Resolved for a few days to sport and play,
And devour the small fishes in the silvery Tay.

William T. McGonagall, 'The Famous TAY WHALE!!!'
Broadside, 15 January 1884

APPEARANCE OF A WHALE AT BROUGHTY FERRY: In the gloaming last night it was seen at the Tayport side of the river, its movements being eagerly watched by crowds of spectators.

Edinburgh Evening News, Tuesday, 13 November 1883

Part II: Whalers take the bait.

When it came to be known a whale was seen in the Tay,
Some men began to talk and to say,
We must try and catch this monster of a whale,
So come on, brave boys, and never say fail.

A WHALE IN THE TAY: During the forenoon the whale was again near the same place, and about half-a-dozen whale boats went out in pursuit … but the big fish did not again appear. The Tayport mussulmen, however, reported that they saw it several times in the shallow water abreast of Luckie Scaup.

Dundee Advertiser, 17 November 1883

DUNDEE SEAL AND WHALE FISHING FLEET: The Dundee seal and whale fishing fleet for next season will consist of seventeen vessels – thirteen steamers and four sailing ships. This is the largest whaling fleet which has ever sailed from Dundee.

Arbroath Guide, 1 December 1883

REAPPEARANCE OF THE WHALE IN THE RIVER: [O]bserved enjoying himself below the Ferry. Several boats manned by seamen belonging to the fleet, under the command of expert harpooners, left the Harbour for the purpose of capturing his lordship …

Dundee Courier, 4 December 1883

Part III: Advantage Whale.

Oh! it was a most fearful and beautiful sight,
To see it lashing the water with its tail all its might,
And making the water ascend like a shower of hail,
With one lash of its ugly and mighty tail.

Then the water did descend on the men in the boats,
Which wet their trousers and also their coats;
But it only made them the more determined to catch the whale,
But the whale shook at them his tail.

EXCITING WHALE CHASE IN THE TAY: He ducked and dived, puffed and spouted about in the highest state satisfaction and hilarity. He was evidently in his element, and enjoying a veritable feast of good things in the way of sprats and herrings. The news spread like wildfire through the Ferry ... his whaleship, all unconscious seemingly of the sensation he was creating ... the 'downy' leviathian seemed to relish the joke immensely leading the whalers a dance, within sight of the very oil tanks yawning for his blubber.

Dundee Advertiser, 7 December 1883

ON THE TRAIL OF THE WHALE: One of the boats was close on the whale, and the harpooner fired his gun, but the whale gave a grin and disappeared below the water.

Dundee Courier, 7 December 1883

CHRISTMAS GREETING FROM THE WHALE: Yesterday morning the whale reappeared in the river below the Horseshoe Buoy. Apparently the present visit of the big fish is to salute his friends the whalers with a Christmas greeting.

Dundee Courier, 25 December 1883

THE WHALE INTERVIEWED BY HIS MOTHER ON HIS EXPLOITS IN THE RIVER TAY: 'Oh! where have you been, my son, my son?/ We have not met since the morn was young.'/ 'I left the North, good mother, to see/ The whaling fleet in bonnie Dundee.'

'Spectator', *Dundee Courier*, 27 December 1883

THE WHALE IN THE TAY: Yesterday afternoon a splendid sight of the monster was obtained at Broughty Ferry. It rose about 100 yards off … and leaped clean out the water. The movement, which was repeated three times, is described similar to the leap of a salmon, with the difference that it seemed a little slower. Rising almost perpendicularly till clear of the water canted on its side as it fell, causing a great commotion in the water, which at the time was as calm as a lake.

Shields Daily News, 31 December 1883

Part IV: The campaign gains a national following.

Then the people together in crowds did run,
Resolved to capture the whale and to have some fun!

A WHALING ADVENTURE: … there were three harpoons in the whale, and it had been struck by several rockets.

Express and Echo, Tuesday, 1 January 1884

THE WHALE HUNT OFF THE TAY. EXTRAORDINARY ADVENTURE: … The harpoons were still in his body when he disappeared, and it is impossible that he can long survive his injuries. Those on board the tug estimate the length of the animal at from forty to sixty feet. He has a very peculiarly shaped head, and is supposed to belong to the 'hunchback' variety.

Edinburgh Evening News, Wednesday, 2 January 1884

THE HUNT AFTER A WHALE: The whale harpooned in the Tay on Monday escaped during the night after being fast for twenty-two hours. Its strength was so great that it towed a steam tug and two boats through the water at a great rate out to sea. Finally the ropes broke, and the monster disappeared.

Bristol Mercury, Wednesday, 2 January 1884

Trigger warning. Some particulars of [the whale's] endurance may be interesting ... Hand-lances were driven three feet deep into it, and blood spouted from the wounds. Two of the harpoon lines parted, but the steam tug and the two rowing boats were dragged out to sea by the remaining line, north to near Montrose, south to near the mouth of the Firth of Forth, then north again. At daylight a four-feet-long iron was fired into it, also a couple of marling-spikes, and a number of iron bolts and nuts. About twenty-one hours after being harpooned it showed signs of exhaustion, turning from side to side and lying level on the water, but shortly revived and again held on; in half an hour the line parted, some way south of the Bell Rock, and the whale was free.

John Struthers, footnote in *Memoir on the Anatomy of the Humpback Whale Megaptera Longimana* (1889)

THE RUNAWAY WHALE: No further intelligence has been received concerning the runaway whale.

Dundee Courier, Friday, 4 January 1884

THE WHALE: The public have grievance against the ancient mariners who came up the Tay on Tuesday forenoon looking somewhat sheepish after having bidden a lingering and reluctant farewell to the whale. His whaleship had become quite a local personage.

Dundee People's Journal, 5 January 1884

Part V: An anticlimactic conclusion.

… first seen by the crew of a Gourdon fishing boat,
Which they thought was a big coble upturned afloat;
But when they drew near they saw it was a whale …

THE TAY WHALE FOUND DEAD: Our Montrose correspondent, telegraphing last night, states that the whale which was harpooned in the Tay last week, and which, after swimming out to sea dragging after it a steam tug and whale boats, eventually broke away from its captors, was discovered yesterday afternoon floating a few miles off Gourdon, on the Kincardineshire coast, quite dead … thus somewhat ingloriously has terminated a career which has been celebrated in song and story.

Dundee Courier, 8 January 1884

Part VI: Sold to the highest bidder.

Then hurrah! for the mighty monster whale,
Which has got seventeen feet four inches from tip to tip of a tail!
Which can be seen for a sixpence or a shilling,
That is to say, if the people all are willing.

THE RECOVERED WHALE AT STONEHAVEN. SALE OF THE MONSTER TO A DUNDEE MAN: The bidding then settled down between Professor Struthers, Aberdeen University, and Mr Charles Ferrier, acting for Mr John Woods, taxman of the Greenmarket, Dundee, to which latter it was, after a very keen competition, knocked down at £226. It is intended for exhibition purposes in Dundee.

Dundee Courier, 11 January 1884

We tried to preserve the heart, but our hands went through it ... The putrid soft parts having been scooped out, and the remaining soft parts prepared with antiseptics, a wooden backbone was introduced, wooden bars supplied the place of ribs, and the body was stuffed and stitched below into proper form. The embalmed whale, thus wonderfully restored in form and much lightened, was exhibited during the next few months in various towns, first in Aberdeen, then in Glasgow, Liverpool, and Manchester, again in Glasgow, in Edinburgh, and finally again in Dundee.

Struthers, ibid.

If the Famous Tay Whale had been left to its own devices rather than hunted to its slow and dishonourable death in early January 1884, then paraded in death around Britain for the entertainment of vast crowds of gawping Victorians – paraded without its flesh, without its organs, without its backbone, its skin stuffed and draped over a wooden frame and stitched up again to look like a caricature of its living self – it might still be alive today.

Jim Crumley, *The Winter Whale* (2008)

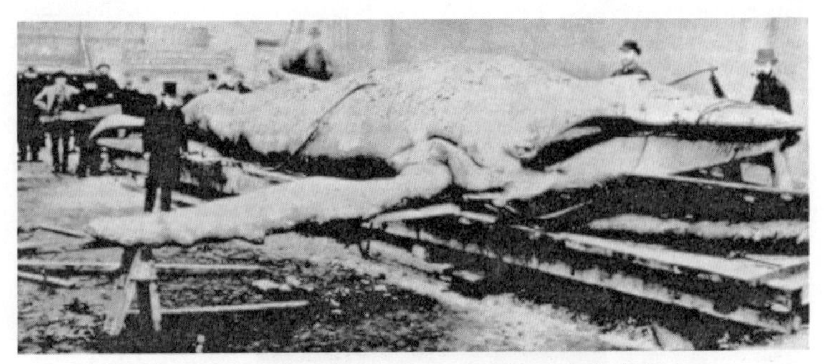

'Wonderfully restored in form and much lightened':
John Struthers (left, top hat) with the Tay Whale
at John Woods' yard, Dundee, 1884.

Postscript: Precarity.

And my opinion is that God sent the whale in time of need,
No matter what other people may think or what is their creed;
I know fishermen in general are often very poor,
And God in His goodness sent it to drive poverty from their door.

Like the fisherman he wrote about, William McGonagall made a scanty and uncertain living, mainly through broadsides and performances. The Tay Whale was also an unexpected opportunity for him. Like the whale, he has been retrospectively reclaimed by the city of Dundee as a 'favourite son'. The now-beloved eccentric dubbed the 'world's best bad poet' died in poverty and was buried in a pauper's grave.

The poet William McGonagall (1830–1902) by William Bradley Lamond (1857–1924).

ALL WASHED UP: Haunting pictures show how Britain's most loved seaside towns have declined over the last thirty years.

Daily Mail, 26 December 2016

'These places feel a bit stranded.'

Ben Pledger, Deputy Director in the Cities and Local Growth Unity at the Ministry of Housing, Communities and Local Government, *Oral Evidence for the Select Committee on Regenerating Seaside Towns and Communities*, 3 July 2018

FALLING OFF A CLIFF. The economic gap between coastal communities and the rest of the country has widened even more since our 2019 report. 'Many coastal areas are still poorer now than they were before the financial crisis in 2007.'

Social Market Foundation study, 22 August 2019

6

Hull and Harris

December 2016. Alicia due at the end of March. The whales and I seemed to have reached an understanding. Some people think in numbers, some in pictures – some in stranded whales. And just as numbers and pictures correspond to an external reality, so too did the whales. The philosopher Donna Haraway once wrote that 'I am who I become with companion species.' It is in our proximity to non-human animals, so often relegated to the barely perceptible periphery, that we articulate the limits and dimensions of the human self. While admittedly not your traditional canine or feline associates, I (and many of the subculture) would nevertheless want to chime in: yes, we are who we become with companion species. Our necro-companion species – the stranded whale.* We also *become* – *became* – in coeval relation with these creatures; developed our identity from this age-old interaction and convergence.

Those who scavenge the body parts of whales apprehend this most intensely of all: they make their interventions so as to prevent these animals being co-opted into a distinct sphere, set apart from us – the pristine 'natural world' so lyrically hawked by the nature industry. Scavengers underscore the basic material reality: the whales are non-returnable, we cannot reverse the stranding. The philosopher Raymond Williams put it best when he wrote that 'we have mixed our

*I'm borrowing the prefix 'necro-' here from Cameroonian philosopher and political theorist Achille Mbembe. See his life-changing, mind-altering work *Necropolitics* (*Politiques de l'inimitié* (2016) in the original French), translated by Steven Corcoran (2019).

labour with the earth, our forces with its forces too deeply to be able to draw back and separate either out.' He goes on to issue a warning: 'if we mentally draw back, if we go on with the singular abstractions, we are spared the effort of looking, in any active way, at the whole complex of social and natural relationships which is at once our product and our activity.' Those who make their cuts and incisions, who feel moved to dismember the corpses of dead whales, acknowledge the irrevocability. Walking up to a body of a whale, saw in hand, speaks to a desire to salvage a token of the self's invisible extension into the non-human underworld. Maybe this is why she – they – salvaged the jaw, the mouthpiece. A symbolic attempt to find a voice in a world that so forcibly denies the material reality of our combined fates. Stranded whales as surrogate family; stranded whales as an embodiment of nationalist melancholy; stranded whales as symptomatic of the failure of capitalist political economy; stranded whales as the victims of austerity. Our thinking with – our material entanglement with – our necro-companion species.

The weekend before Christmas I received a communication from Big Blue. Mick and Ron had been as good as their word. I knew it would be tantamount to opening Pandora's Box, but I opened it all the same. He had a suggestion for me – a lead. I need to go back again to properly explain it all, to where I left off with Clive and the gang.

And so we unavoidably arrive at the necro-cetacean Mariana Trench: Big Blue. He's dedicated decades of his life to dead whales; their collection, preservation, fashioning, hawking. Access comes with a very serious warning: *no one* traces him.

It had been back in September 2011. I'd found out about him by way of a circuitous Redman Retreat; reached out because I was approaching a period of unemployment. Worth a try. Better than most jobs on offer. With a ropey, half-finished PhD on Herman Melville, it felt as though interfering with the corpses of whales was the kind of labour I was most qualified to perform. I'd got over the relationship, shovelled it into landfill anyway, but was now in fairly dire financial straits. My PhD funding had run out and I was cobbling together bits and pieces of work trying to stay afloat.

The mainstay was an outfit above Lloyd's Bank in Cambridge. Dr Kim ran a school for Korean kids taking their GCSEs and A-Levels. Kim paid me £15 an hour and charged them triple that. I was truly grateful for the work, intermittent though it was. My first week, and a businessman marched his eight-year-old daughter into the room. 'Teach her Shakespeare.' He left the room. The little girl looked at me and said: 'I'd like to start with *Romeo and Juliet*, please.'

Big Blue's first communication had read: '*Whale in field. Fin? Skeffling. Equinox. Too late but worth a look? Meet Saturday. Address to follow. I'm sure we can find something for you to be getting on with.*'

On a Thursday night I got into my car and drove three and a half hours north, hoping the investment in petrol would pay off. I slept in a service station outside Nottingham, a Big Mac come-down easing me into sleep. I passed through Hull at dawn. Skeffling is half an hour further along the Humber Estuary. A population of 150 bisected by the Greenwich Prime Zero meridian line. Straight north from London, across the water from the gnarly fishing port of

Grimsby. I parked at the parish church and asked someone about the whale. I looked pale. A woman in a green felt hat pointed towards the tidal marshes. I put on my mud-covered Dunlop boots and struck out towards the sky-blue estuary. Not a breath of wind, and hives of sparrows argued in the autumn hedgerows.

The whale lay slumped half a kilometre from the shoreline in a recently harvested field. A juvenile sei (not fin) some thirty foot long. Inexperienced and looking for food, it had been caught out by that year's equinox tide. An alignment of Sun and Moon (instead of just Moon) momentarily exaggerates the gravitational bulge through which the turning earth passes. The resulting tidal surge allows people to surf up Britain's major rivers, and this creature had nearly trespassed into the village. Equinox whales get left in the most incongruous places. Every March and September, you are likely to see a strandings enthusiast stalking an early morning Humber, Thames or Severn tideline. The excess always leaves something behind. One humpback whale was recently found deep inside the Amazon rainforest, having accepted a particularly wild inland ride.*

The sei whale lay pointed back towards the lee of Spurn Bight, a stretch of water sheltered from the open ocean by the extended north bank of the Humber. Beyond that, Denmark. Realising its mistake, the whale had tried to turn itself around and swim back. Its mouth was now open, tongue beginning to bulge, skin tightening to a familiar

* See Matthew Haag, 'Humpback Whale Washes Ashore in Amazon River, Baffling Scientists in Brazil,' *New York Times*, 25 February 2019.

synthetic rubber sheen. Ought I to try to retrieve some hide as a sign of my commitment? No. People were coming. The colours of the Yorkshire Wildlife Trust. I got up, drew in a lungful of air, and walked slowly back to my car. The kind of beautiful hazy day English autumns are sometimes capable of. I woke up at noon, the faces of two men I didn't know staring in through the windshield. They were mouthing something about parking. I cracked open the door and said in my politest accent that I was about to leave. One of them handed me my keys and I thanked him.

There was a message on my phone from Dr Kim, asking whether I was available to give immediate feedback on someone's A-S Level coursework. £35 for comments and a two-hour-long phone call. He'd charge a hundred and I'd have to subtract the internet café.

My plan was to sleep in the car on Friday night, somewhere overlooking the Estuary, maybe at Paull. I'd spend Saturday day at the Hull Maritime Museum, catching up with old exhibits. It was 1.30 by the time I managed to find a parking space in Hull city centre; 2.45 when I sat down to a computer, in an empty room lit by a buzzing neon lamp in spite of the blazing sunshine outside. I opened the file. Decent work. The kid on the other end of the line was knowledgeable and unhappy; didn't need a two-hour conversation with me. At his whispered request, we spent some of the time in silence.

Another message: '*Meet at 2. Near the scrimshaw.*'

Saturday lunchtime and I was standing in front of the scrimshaw collection at the Maritime Museum. There's nothing like it outside the United States. Etched portrait-wise onto the base of my favourite tooth is a lowered whaling boat. A harpooner, standing on the prow, has just conquered a sperm whale. In the foreground, the whale gasps upside down at the sky while the mother ship waits patiently in the background. Its mainmast doesn't even reach the mid-point of the tooth canvas; instead it gestures upwards towards an expanse of swirling, discoloured ivory enamel that seamlessly transfuses an unmarked horizon into a tapering sky.

Two people entered the room. I turned around and nodded in half-greeting, just in case. A man with a full head of white hair gave a small acknowledgement; a woman, a few years younger beside him, slightly lop-sided with the weight of a bulky satchel. They approached the display case.

We stood together in silence and the man began: 'Two years ago, an elderly lady from Macclesfield was rooting around at the back of her wardrobe, and happened upon the portraits of three young adults and a child.' Turning to me. 'Did you hear about this?'

My blue eyes met his blue eyes. 'No, I don't think so.'

'They'd been abducted in 1830 from the tip of South America by the captain of the HMS *Beagle*, Robert FitzRoy – FitzRoy is a distant relation of mine.'

I made a small noise.

Turning back to the display case, he continued: 'On the *Beagle*'s first voyage home, the abductees were carved onto the side of a sperm whale tooth. The work of Royal Marine and scrimshander James Adolphus Bute. Have you heard of Bute?'

I had.

'It was one of several works he produced during the *Beagle*'s voyage. Bute gave the tooth to his friend Private Thomas Burgess, who later became a Macclesfield policeman. That's how the portraits ended up in the family wardrobe a hundred and seventy years later. Adam Partridge Auctioneers and Valuers put an estimate on it of five to ten thousand pounds.'

Silence.

I asked him if he'd bought it.

'Partridge sold it for six thousand pounds. To a "mystery buyer". A few months later, the same tooth went up for auction at Bonham's London. It exceeded its estimate of fifty thousand pounds.' He gestured to his partner. She opened the satchel and passed him the magazine. He turned to a particular page and showed me the photo of Lot Number 6135.

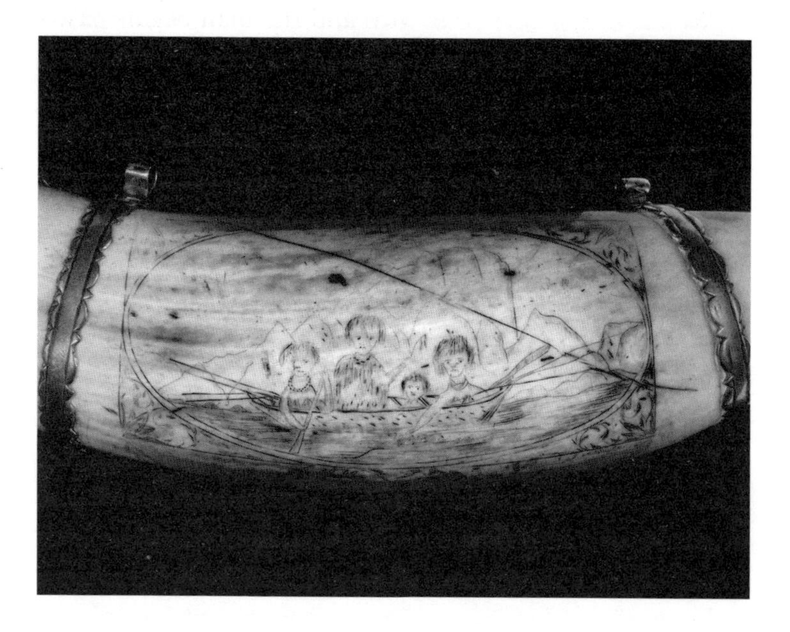

Pointing with his signet-ringed little finger: 'Four people, stolen from the various communities that populated Tierra del Fuego, the southernmost tip of South America. The second from the left is most likely a young man from the Alakaluf tribe called Elleparu. That was his *Yahgan* name. The crew renamed him "York Minster" because of his size. On the left holding a paddle: that's Orundellico, later "Jemmy Button" because he was paid for with a mother-of-pearl button. He was a favourite of the crew. "Boat Memory" is on the right. His name is lost.'

'Who's the child?'

'She is an eight-year old called Yokcushlu, renamed "Fuegia Basket" because of her basket-making skills.'

He handed the magazine back to his partner. She put it back in the satchel and pulled out a book: Volume 1 of *Narrative of the Surveying Voyages of his Majesty's Ships Adventure and Beagle between the Years 1826 and 1836, Describing their Examination of the Southern Shores of South America and the Beagle's Circumnavigation of the Globe*. An 1838 first edition. He turned to a book-marked page (416): Yokcushlu 'was so merry and happy, that I do not think she would willingly have quitted us'.

He looked at me for a response.

Continued: 'Natives of Tierra del Fuego, better suited for the purpose of instruction, and for giving, as well as receiving information, could not, I think, have been found.'

He closed the book and gave it back to his partner. He looked at me again.

'They captured them,' he continued, 'and James Bute captured them again on the tooth for posterity. On the voyage back to England, the *Beagle* stopped at Montevideo. All four were vaccinated against smallpox. No-Name-Boat-Memory died of smallpox in Plymouth. In 1831, FitzRoy placed the others with a schoolmaster and his wife in

Walthamstow. They were paraded at Court and met Queen Adelaide. She had only been able to give birth to the still and so pronounced Yokcushlu a surrogate daughter, gave her a pretty bonnet. She was charmed by the little girl's English greeting of "How do?"'

'"How do?"' I said, stupidly.

The man nodded.

'FitzRoy decided that the three survivors should sail back to Tierra del Fuego a year later. That was the *Beagle*'s second voyage; Charles Darwin was on board that time – he was well acquainted with these people.'

'Thank you. Fascinating', I said.

'Yokcushlu and Elleparu married. Yokcushlu married again after her husband died in a knife fight over a property dispute. She died in 1883.'

Another small noise of appreciation. His partner stared blankly at the display case.

He continued: 'Orundellico – little Jemmy – led the massacre of Christian missionaries at Bahía Wulaia Bay in 1859. Opposite that bay is an island called Button. It's named after him.'

He stopped, measured a pause. Looked at me. The ticking of a grandfather clock.

'What was it – fin or sei?'

Sei.

Big Blue insisted we drive in their car. We'd be back by early evening. His partner did the driving and I sat in the back with him. Executive leather seats from the eighties. I should know he was related to the Royal Family. No doubt I'd noticed the signet ring. His late father was a baron. Mother was the daughter of his steward. Everything was hushed up.

He made sure they were provided for; gave Blue some money when his mother passed. He could afford it, too – the family had made a series of investments in late-eighteenth-century sugar; were by all accounts handsomely compensated by the British government in 1833 when everything was settled.

He looked out of the window. 'A slave-money trustafarian.'

He delivered the last line as a joke, and an anger momentarily rippled through his expression when I didn't smile.

An awkward fifteen-minute drive through gradients of English farmland. Somewhere between York and Hull, a leafy town and a detached Victorian house. Three floors held up by a matted carapace of ancient ivy. Three living inhabitants: two humans, one tortoise. Allegedly, the tortoise was as old as the house. For the entirety of our meeting, it sat in the corner of the dining room, looking hopefully up and out through the ivy-covered window – mouth slightly ajar. Big Blue intermittently scattered salad leaves in front of it which it studiously ignored. They hadn't decided on a name, had purchased it in the seventies; wanted to work with fresh tortoise shell at the time and found they didn't have the heart to dispatch it.

'You know the French were very cruel to tortoises and turtles? Do you know what they did, the French?' He moved over to the bookshelf that covered the entire wall and took down Auguste Escoffier's 1907 *Guide to Modern Cookery*. Turning to a particular page, he read: '"*The Slaughtering of the Turtle*. For soup, take a turtle weighting from a hundred and twenty to a hundred and eighty pounds."'

'As big as I am', his partner interjected.

'As big as Louise.'

He continued, '"Let it be very fleshy and full of life. To slaughter it, lay it on its back on a table, with head hanging over the side. By means of a double butcher's hook, one spike of which is thrust into the turtle's lower jaw, while the other suspends an adequately heavy weight, make the animal hold its head back; then, with all possible dispatch, sever the head from the body."'

For most of the conversation, Louise, who I understood to be Big Blue's long-term partner, stared through the same window as the tortoise, also with her mouth slightly open. Every now and then, she would get up, leave the room, and bring us all something. Big Blue allowed that, for most of their relationship, they'd both managed 'a sustainable Ritalin dependency'. He didn't touch the stuff any more; she'd gone back to it recently. Hence her state of 'distant focus'. I glanced at the tortoise.

The light of another beautiful autumn day filtered through the window. Louise placed a tea tray on a coffee table made of the shoulder blade of a North Atlantic right whale. They'd found it during a family road trip to Norway; had acquired most of their bones up there, though further north – even to Svalbard – was best, apparently. 'The sea ivory lies in great heaps at those old whaling stations.' During the early 1980s they'd brought back substantial quantities; crossed between Bergen and Newcastle. Customs never searched their van, or if they did, they were given a sufficient amount of money to stay quiet about it. Security only really tightened up in the late 1990s. Everywhere in this room, decaying shades of ivory whiteness.

Big Blue watched me as I looked, asked if I wanted him to talk about anything in particular. I took my time. The most breathtaking private collection of whale bones I'd ever seen. One item immediately caught my eye – it was meant to: a giant 'Dun Cow rib' leaned against the bookcase and curved back into the room. A lamp, fastened to its tip, illuminated a lump of granite that had been positioned on a driftwood side table.

Before I asked the question, he'd started talking. 'That's an important piece. Would you like to try some?'

I said I wasn't sure what he meant.

'It has very particular properties. I have some prepared. A small quantity in your tea is best, instead of sugar.' From a drawer, he took out a small vial of dust, carefully opened it, and dabbed a few sprinkles into each of our cups. Louise picked up the teapot and began pouring. 'It won't do you any harm.'

'This is granite?'

Big Blue nodded.

I brought the cup to my mouth, looked at them both, and sipped my tea.

'So what properties does it have?'

'I'm not going to tell you yet.'

'Have you just drugged me?' I said, politely.

'No, nothing like that.'

The next silence lasted minutes. I fixed my stare at the intact juvenile sperm whale jaw rested against the back of a faded crimson chaise longue. I began this time, wanting to show my stripes. 'I've been trying to track down a piece recently. You'll know about it. The chair that's mentioned in *Seals and Whales of the British Seas*? Included by Redman, of course.' I took out my latest notebook. 'Thomas Southwell describes the basal portion of a skull of a sperm whale which has been converted into a chair. In the Church of St Nicholas at Great Yarmouth.'

He went over to his bookshelf, flipped through an album, and took out a postcard. 'This one?'

He asked if I'd care to share my notes.

From my black notebook I read: 'Fifty-one-foot sperm whale washed ashore at Caister, 1582. At some point during decomposition, a parishioner identified the cranial seat and first vertebra as chair-like: four foot six inches wide, and four foot high. By 1606 it had been installed somewhere near the entrance of the churchyard where, to the increasing unease of local clergy, it began developing a reputation for sorcery and divination. "An object of wonder relegated to the powers of darkness", Southwell writes. On their wedding day, couples were encouraged to dash to what was now known as "The Devil's Seat": whoever managed to sit on it first, so the lore went, would thereafter wear the trousers in the relationship.'

I looked up. He nodded encouragement. 'By the early nineteenth century, it's likely that a local vicar used a renovation of the churchyard as an excuse to quietly retire the Devil's Seat from its ritual function. However, he didn't dispose of it; he "admitted it into mother church", where it stood for many years "beside the north-west door under the clock". During the bombing of 1943, it was badly damaged and (possibly?) thrown out. Someone must have salvaged it, though.' I looked up at Blue.

'They certainly did. Now the throne of a chap who lives just outside of Droitwich. A question for you: why do you think the priest brought a Devil's Seat into the church?'

I paused. 'I'm not sure – he wanted to rehabilitate an important community relic?'

Big Blue seemed pleased. 'No – the priest saw something of what its original salvager saw. The chair wasn't just an allusion to the Old Testament; it was the precise spot where Jonah sat and waited for absolution. It was positioned at the north-west entrance: the sinners of Great Yarmouth were invited to sit down, as Jonah, look up at the church's vaulting ribbed ceiling, and pray (from inside the whale) for God's mercy: "Out of the belly of hell cried I, and thou heardest my voice."'

Big Blue picked up his tea and sipped. Said all of this with a satisfied 'You won't find that in your Redman' air. He glanced at Louise. She looked bored by the exchange, but took this pause as a cue.

'It's come to our attention that there's a sperm whale currently rotting on Taransay. Are you familiar with the Outer Hebrides? Our contact there is recently retired.'

I nodded and asked if they had many of these contacts. Blue explained that once upon a time, an informal network covered every region of the coastline: clockwise from Portland in the South, along and around to Plymouth, Lundy, Irish Sea, Malin, Hebrides, Fair Isle, Cromarty, Forth, Tyne, Humber, Thames, Dover – and back full-circle to Wight. It had now whittled down to just Humber and Thames on the east coast, an area covered by a pair of ancient brothers, apparently. The fastest, most discreet tooth extractors in the east.

'You can be our man in the Hebrides. On a trial basis. Fetch us this whale's whitest teeth. We'll give you the petrol and ferry too.'

I sat considering this for a moment. When would they want them by?

As soon as possible.

I decided to lean in.

'All right.'

We sipped our teas. I continued to look around the room. 'Can I ask why you want more teeth? You seem to have quite a lot here.'

'Some customers want them white as possible and fresh. They find the yellowing enamel of antique ivory off-putting. They prefer non-smokers.'

I smiled politely. 'Who are your customers?'

"The Chinese market, mainly. As I say, the premium's on fresh produce. That's been our mainstay these past twenty or so years. We've had other requests. Dinner tables, office chairs – archways. The market has changed. New laws and regulations have created a very lively black market; they land with spiralling prices on their heads.'

More tea.

Two years ago, a Swedish man had got in touch about the possibility of fashioning an assortment of sex toys: three butt-plugs, three dildos. He'd been inspired by some objects he'd seen at Oliver Hoare's Cabinet of Curiosities in South Ken. Had I seen those?

I hadn't.

'You must. He's planning an exhibition. A very curious man.'

I sipped my tea.

'The Swede – we called him "Andromeda" – wanted his items bleach-white and textured. I reminded him that the butt-plugs would have to be fashioned from whalebone – sperm whale teeth are hollow – and it would otherwise be impossible to get the desired shape. But this was of course an advantage because he'd get the roughness of the bone. A man interested in texture would also prefer switching the dildos from teeth to bone, but he was adamant about the teeth; wanted to simulate the experience of being "bitten down into". It was all relatively straightforward in the end: we just jigged slight ridges onto the surface of the enamel, steeped them in our own recipe of wax sealant for hygiene's sake and they were ready to go. Delivered them to a PO Box in Southampton. They're on board his yacht now, released back into the wild.'

'Where do you do the work?' I asked.

'The garden shed. You're not going to see that. It's a mess.'

'I don't mind mess.'

'Not today', said Louise calmly.

We sat in silence again. Big Blue walked over to the

bookshelf. 'You ought to give me your address.' He picked up a Filofax, turned to a particular page, pen at the ready. Still holding out for access to the workshop, I gave it to him. He finished jotting it down, turned to another page bookmarked by a photograph. Lingered there thoughtfully, and then, before I knew what was happening: 'I feel I need to tell you about the Queen Mother.'

I didn't say anything.

'A sorry case.' He looked concerned. 'She's remembered for the fly-fishing. A really keen fly-fisherwoman, wasn't she? Everyone knows that. But what they don't tell you is that she didn't return a single fish she caught to the water.'

He continued. 'In the nineties she almost choked to death on a salmon bone. Told the press that it was "the salmon's revenge". The newspapers reported this as Queen Motherly wit. It wasn't. She was being deadly serious.'

I continued to sit quietly.

'What we're about to tell you has been buried. Along with everything else. It pains me to say it but you can't keep something like this hidden for ever.'

A deep, regretful sigh.

'They had a terrible time with her at Glamis Castle. The governess started raising concerns as early as five. It wasn't her fault – apparently she'd been privy to a particularly violent shoot, an afternoon of hundreds of birds raining down from the skies, exploding in clouds of blood above her head. She left the field drenched in gore, cuddling a teddy bear.'

Louise joined in: 'A sorry business. Some say she was inspired by that two-headed sheep they keep at the Bowes Museum. Dreadful thing.'

With some excitement in his voice now. 'She started stalking the grounds of the castle, looking for animals. It's why they're so protective of the archive – because of her. At seven she was a nightmare. Leaving ghastly hybrids around the house and gardens: the notorious "animal kebabs" of Glamis Castle, they called them – the head of a pheasant stuck onto the body of a toad, a robin with its legs pulled off and placed in an egg cup. At this stage, some in the family still thought this an expression and extension of her precocity.'

'She was a very talented classicist.'

'But then it progressed. The severed head of a much beloved hunting dog with a lamb for a body; another wearing its own intestines as a scarf. She left that particular totem in the family rose garden. Her youngest brother, the five-year-old Hon. Sir David Bowes-Lyon, was so frightened he refused to leave his bedroom for a year.'

He removed the bookmark photographs. One, a magnified version of the other. 'Look at her.'

He moved over to a side table, opened it, took out a small wooden box and raised it to his nose; snorted a significant amount of something into his right nostril. He offered some to me. I declined. Then into his left. Eyes watering, he moved over to the tortoise, anxiously removed some of the more wilted salad leaves. He continued, whispered British conspiracy in his voice. 'There were projectiles, too. She'd slaughter something – a fawn, or a cat – take the carcasses to the top of the castle battlements, and then hurl them down at anyone unlucky enough to be standing in the courtyard below. She allegedly struck the visiting Prime Minister Sir Henry Campbell-Bannerman with a still-born deer.'

'He was very gracious about it,' Louise interjected. 'But it was clear to everyone that something had to be done.'

'They tried barring Elizabeth from the upper levels of the castle. She was very firmly rebuked by her mother and father. Some weeks passed. She was nine. The family hoped she was over the worst of it. Then a gamekeeper was hit by a flying piglet with such ferocity that he lost an eye. Somehow, she'd managed to construct a catapult large enough to send small- to medium-sized animals hundreds of yards into the air. The family finally acted. She was confined to her rooms – made to promise she'd stop, with the threat of sending her abroad to an institution.'

'Did she stop?' I asked.

'A local fisherman showed her a way of channelling her passions into salmon and trout fishing. Cured her.' Big Blue blinked. 'There were further flare-ups of course, even in later life. But mostly she managed to pass for a particularly passionate blood-sports woman.'

A lull followed by another surge.

'Have you seen ever seen a Punch and Judy show? The

sort of thing they used to have at summer fêtes? You're too young.' I confirmed that I'd seen a Punch and Judy show. 'Every charity event she organised, she insisted on putting on a Punch and Judy show. Her twist was to replace the crocodile with the gaping, hastily butchered head of a sea trout on a stick.'

'I had no idea.'

His eyes were all over the place. I was becoming anxious. Couldn't help looking down at the tea. 'Would you mind telling me about this now?'

'You're an impatient young man, aren't you?' Louise said.

'I'm interested.'

Big Blue started talking. Louise cut across him. 'It's a piece of English granite. We found it on a beach in Ghana.' She let that sink in.

'English granite, Ghanaian beach. Why do you think it was there?' I stared at her.

'Another relation of his was Governor of the White Castle. Cape Coast Castle. Britain sailed its ships from Bristol, Liverpool, Glasgow. Weighed down by rocks like this one. Unloaded them to make room for the cargo they wanted to sail across the Atlantic. The beach was littered with the stuff.' She looked out into the light.

'You've made me drink the ballast of a slave ship?'

'It keeps us on an even keel,' said Louise blankly.

I looked at her; looked at Blue. 'What?'

Big Blue got up, reached into his pocket, and counted out ten fifty-pound notes. 'Let me know if you find anything.' I tried to say something; said nothing. Took the money and got the fuck out of there.

*

I got home, did my teaching for Kim, packed a tent, some tools and supplies, and by Friday I was on a train headed for the west coast of Scotland. I arrived at Oban ferry terminal, bought a bottle of whisky, boarded the boat, and sat on deck. Against an Atlantic-Hebridean storm front, sun-white gannets dive-bombed the ferry's wake. About two hours into the journey, the captain announced that a whale-shark was swimming fifty metres to starboard. I stayed put. The buses were on time and I had to take four of them. South Uist, Benbecula, North Uist – another ferry across to Harris. It was raining so hard when we docked at Leverburgh I was forced to check into the hostel by the quay. I shared a bunkroom with a massive windsurfer called Rory. We had breakfast together. He pointed out that the wall at the front of the property was the spine of a fin whale. Ten giant vertebrae: the person who gave us eggs said they'd only recently stopped stinking the place out.

In the morning sunshine, Rory took me as far as Seilebost, the uninhabited Island of Taransay arching up out of the middle distance. I thanked him. He joked that it almost looked swimmable; wished me luck. After an hour's walking, I met a likely someone, asked how I'd go about getting across. He must have been in his early fifties. He said I could ride out with him on the afternoon tide if I liked. He wouldn't accept my money – was going out there anyway – he'd pick me up the following day. Did he know about a sperm whale? Surely I could offer him something? No and no.

I walked Taransay's perimeter in a couple of hours. Nothing. At 8 p.m. it got cold. I found a leeward beach, collected a stack of driftwood, doused it with the small bottle

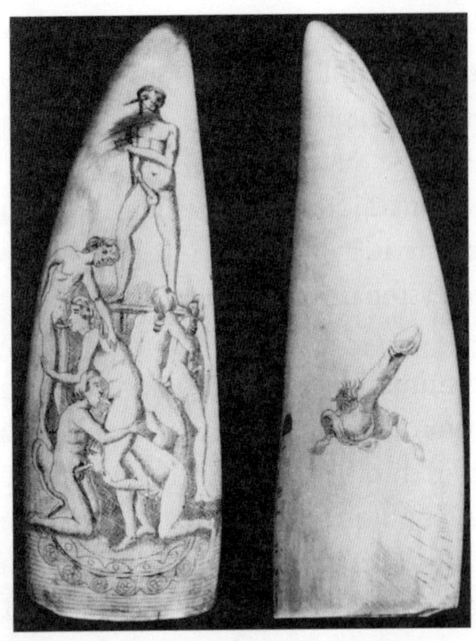

Catalogue Entry: XX// A. SAILORS AT PLAY// B. A FLYING PHALLUS// *Sperm whale tooth// Size: 15cm long* (pp. 94–95)

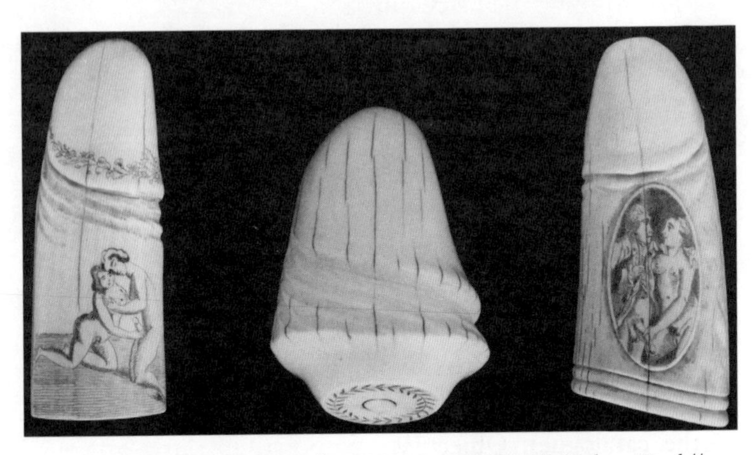

Catalogue Entry: XXXI// LOVING COUPLE// *Walrus Tusk// Size: 10cm long*// XXXII// THE TIP//*Walrus tusk//Size 7cm long*// XXXIII// A STIMULATING ENCOUNTER// *Sperm whale tooth//Size: 12cm long* (pp. 98–99)

of kerosene I'd bought along with me. I opened a tin of beans, ate, and dozed. I woke up freezing cold, went over to the smouldering fire, picked up a warmed-through rock and held it until sunrise. Next day, a second search. Nothing.

My message read: 'I didn't find anything. I'm sorry.'

A real crossroads in my career. The moment I didn't become a professional whale scavenger.

Blue's next communication, four years later, was in the form of an exhibition catalogue: Oliver Hoare's 'Every Object Tells a Story' exhibition in Fitzroy Square, London. Two fluorescent sticky notes bookmarked the main attraction: a collection of forty erotic scrimshaw dated 1800–1900. Thirty-one sperm whale teeth, two whale ear bones, and seven walrus tusks.

An accompanying note read:

Dear P,

We hope you are well. Just in case you missed this. Two objects stand out here: V and VI. Both a cause for curatorial anxiety, and for good reason. Look at Object VI. One side is labelled 'THE CAPTAIN'S ENTERTAINMENT':

Hoare decided not to display the reverse – the 'PLAN OF A SLAVE SHIP' (you'll see it's been censored in the catalogue). Object V, another tooth (and again censored) comprises: A. 'THE CAPTAIN'S PERKS' and B. 'PLAN OF A SLAVE SHIP'. Hoare notes the 'link between whalers in the Arctic and slavers in the

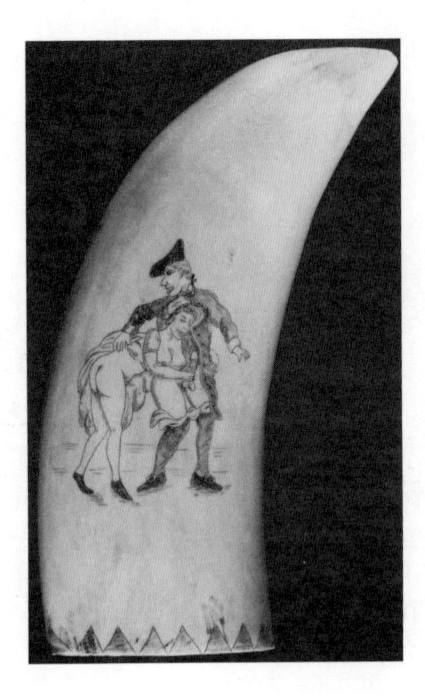

Caribbean' (p. 80). Sometimes we don't know whether we're surrounded by the bones of humans or cetaceans. The Atlantic is wide and deep. May your curiosity be the bowsprit of the good ship of your experience in life. Come back and visit again one day.

L & BB

P.S. It's very good to hear from you, albeit from a third party. Two avenues of potential exploration. Try Mrs Bones. She once stole a jaw. Telephone: –. And talk to Tom in Harwich, who knows about this sort of thing. Telephone: –

P.P.S. 'Look at her in this one!'

7

Harwich and Cromer

My initial response to this communication was to pathologise it in precisely the terms I'd initially pathologised that first insane meeting with them. The confusions, historical elisions – conspiracy theories – were the result of long-standing drug addiction and a life spent floating and imagining on a cloud of inherited money. What happens to people when they don't have to work for a living. Dead whales, sex toys, granite, the Queen Mother, pets, animal torture, tea, sugar, the transatlantic slave trade. A depraved and quintessentially British combination that spoke to the depth of their psychological disturbance.

I was only able to hang on to this indignation for a day or two. This was how I'd first conceptualised my own whale-related symptoms. Automatically attributed them to individual paranoia and delusion rather than as symptomatic of something bigger, beyond any particular atomised psychosis. The 'link between whalers in the Arctic and slavers in the Caribbean' … *The link*. Both sets of early accumulators, whalers and slavers, assumed the non-human status of their respective quarries, both forms of violence were predicated on a default supremacy and the right to consume everything. The capitalised blubber of a billion dematerialised whales, combining with the capitalised blood of a billion stolen human lives. Britain's inheritance. The jettisoned ballast that keeps on washing up, that won't be erased. J. M. W. Turner's painting 'Slave Ship' (1840), later titled 'Slavers Throwing overboard the Dead and Dying—Typhoon coming' figures it more or less exactly. Look closely enough in the bottom right of the image and you see manacles, limbs, fish rolled indistinguishably into one another. Our collective anthropo-ichthyic nightmare. The poet Claudia Rankine helped me see this:*

I'm not sure whether Jonah's whale (top right) rushes in to join a feeding frenzy, or stares on in open-jawed horror at new market realities. *The blast blasted blubber beyond all believable bounds.* Is that a spirit fluke rising up out of the ocean, just above the manacled human leg? Or is it the giant tooth of a sperm whale tapering into the sunrise? Or sunset? And at the top there: is that dry land, or the arching back of a sperm whale coming up for air, preparing for a deep dive? Big Blue and Louise had apparently convinced themselves that their life's work was subject to an occult transference:

* Claudia Rankine, *Citizen: An American Lyric* (Penguin, 2014).

cetacean remains morphing into human remains. Easily done. The shoulder bones of the Saints, the rib bones of mythic English giants – why not the human cargo transported across the Atlantic? The backbone of Empire rearing up out of the ocean, slumping across our consciousness, refusing its own disposal. History. Mistaken as nature.

I tried Mrs Bones first. She said she couldn't help me, but I'd be welcome to pay them a visit if I was in the area. Then Tom. Blue had already been in touch with him and he'd asked around, said there was a likely candidate. A couple of years after the time I'd mentioned, early 2000s, a young woman had approached a dealer friend of his specialising in antique scrimshaw. She'd offered him a sawn-off jaw; £40 he'd given her on account of its missing teeth. He'd picked it up from a house in Harwich. Did he remember which one? Tom had an address for me.

'Thank you.'

Did he know what happened to the jaw?

'Shifted fairly quickly', was his immediate answer. These things were easier to move back in those days.

Did he ever see her again?

No, a one-off.

I was thinking of driving up in the next couple of days, suggested we have a coffee.

'Best not.' Said he'd done right by Blue and basically put the phone down.

I took off east, past Stonehenge in the fading light. I arrived at the M3/M25 intersection on the outskirts of west London too tired to carry on. I parked somewhere, Sunbury I think it was, and slept; woke in the small hours and got myself to Harwich. I stood in front of a terraced house. The curtains were drawn and a stiff onshore breeze scraped a snapped television aerial along some broken guttering. Juvenile seagulls called from every chimney pot. Above me the low rumble of an aeroplane, its banking form just visible among the clouds.

I knocked. No answer. Knocked again.

Two figures emerged from a house fifty metres down the road: a very old man and a near-dead German shepherd; started slowly walking down the road towards me. Knocked again.

When they were close enough: 'Excuse me.'

Neither the man nor the dog heard this.

'Excuse me.' Quite loud now.

He looked up at a second-floor window, down the street, and then back over his shoulder at me.

'Hi – apologies. You don't happen to know who lives here, do you?'

The man moved his lips in tandem with mine. Then looked the building up and down. 'Who lives here?' he said.

I nodded. 'I wondered if you knew.'

'No idea.' A moment of pause while he examined me. 'They're probably at work,' he said.

He and the dog carried on their way, the breeze scraping a polystyrene burger box along after them, nestling in one of the potholes that pockmarked the road. A woman in high-street chemist overalls opened the door.

I was looking for a long-lost relative. A matter of inheritance. You didn't happen to live here in the early noughties, did you?

They'd bought it two years ago. Old lady lived there

before them, though – and for a long time, judging by the décor when they moved in.

'Do you have a forwarding address by any chance?'

'I'll write it down for you, love.'

Ms ____. Sunnymead Nursing Home.

'Could I speak to ___, please? I think she's a resident of yours?' An old friend of her daughter's – I'd very much appreciate a chat and a catch-up.

The voice on the other end sounded surprised I'd called.

At reception an agency nurse called Steve greeted me, no older than twenty. I asked how she was doing and handed over a pot of pink and red geraniums. Steve led me down a corridor that smelled strongly of disinfectant and mushroom soup. 'How long has she been here?' Steve stopped at the door, knocked and opened it without waiting for an answer.

'You're the first visitor in a while.'

Two beds in a blank room, one woman lay on her back staring at the ceiling, mouth wide open. On the little table by her bed there was a plastic cup of water. Her roommate lay sleeping, back turned away from us and facing the wall. She was wearing a beanie. On her table, a couple of framed photographs. With gentleness in his voice, Steve asked if I would mind waiting outside. 'Peter's here to see you. Like we said. A friend of Melissa's.'

Steve reappeared after a minute or so, nodded encouragement.

I walked into the room with a smile plastered across my face and courage draining away with each step. She half looked at me. Her face was drawn, and I realised I'd never once had to step into a place like this before.

One of the photographs. It was her, looking right at

me, uncertain smile meeting the gaze of the camera, different to the face I remembered. I stayed for twenty minutes, intermittently filling the silence with platitudes; said that Melissa had been an inspiration to me. Her mum managed a gulped and whispered goodbye. I said I'd come back and visit again, asked Steve if he wouldn't mind passing on my contact information. He'd be glad to.

On the way out he said that Melissa had never been in touch. Lived abroad. Where? Data protection. I thanked him – wrote my email in my notebook – wrote underneath: 'the boy on the Norfolk beach'. Handed it over.

And then I paused, returned to the desk, asked for it back. Said it was maybe better to leave it for now. Put the piece of paper back in my pocket and said my goodbyes.

Out of courtesy and curiosity I paid a visit to Mrs Bones. My last trip before becoming a father. On the final stretch of country lane, a barn owl wheeled bright as an angel across my windscreen. 'Mrs', because she was still technically married; Bones because of the bones. She lived along one of those stretches of East Anglian coastal desolation. A substantial farmhouse held in trust by several women who'd all agreed to pass it on to someone suitable when the time came. I met two other residents, Aggers and Cat, who also identified as witches. Aggers was shortened from Agatha, an affectionate allusion to the former England fast bowler and *Test Match Special* commentator Jonathan Agnew (though Aggers was considering a change because the programme had lurched so dramatically to the right in recent years, particularly, they said, with the introduction of Michael Vaughan and Graham Swann, who could often be 'real bastards'). Cat – that was just her name.

Aggers was in the front garden tending a raised bed in a knitted brown woollen dress. She looked at me, nodded, and disappeared around a wall. A few moments later Mrs Bones appeared, dressed in a grey smock and felt hat. 'You've come.' She rearranged the position of a birdfeeder as she approached. She led me through the garden gate, up the garden path, and stopped me at the doorway – took from her pocket a bunch of dried sage leaves, and lit them as you would a cigar, flicking the hot embers at me and blowing smoke into my eyes.

She walked through the door and I followed her into a living room. Three whales' bones: two vertebrae and the (quite badly flaking) jawbone of a Greenland right whale. She sat down on one of the vertebrae, placed the sage into what consequently looked like a smoking cauldron on the other. The jaw leaned securely in the corner of the room. She told me to sit, indicating a rug in front of her. A gentle draught from the window played with the hem of the smock. I sat.

'We didn't find who you're looking for,' said Bones.

I told her who I'd just met, what I'd found out.

Sounded like a lone wolf to them: 'The operation hadn't been meticulously planned, had it? Conscripting the help of a young boy.' As far as motives went: the comet, a young woman, a comet tattoo, a whale jaw – three years before the new millennium. A very particular constellation. They were all preoccupied with the Apocalypse back then – augury, divination. If Mrs Bones were to put money on it, the blue-haired comet woman called upon the jaw (maybe individual teeth) to answer some questions. A rhabdomancer – using the severed jaw of a sperm whale (in place of a traditional rod or wand) as a means of divination. 'You'd mark out a circle or pentagram on level ground, charge particular areas of the landing zone as either positive or negative, and then begin casting. See where it fell.'

There were more disconcerting possibilities: the missing teeth – the magic of dragon's fangs, sown in the hope of raising the dead. An army of the dead. As revengeful Cadmus had done under the guidance of Athena. What had she been trying to resurrect?

I said I'd chosen to be at peace with the possibilities.

A few moments' pause while the sage continued to burn.

'I'd be interested in these', said Mrs Bones, indicating the vertebra she was sitting on and its companion. She'd acquired them some years ago – that's how they knew Blue. A female fin whale. Washed up in the eighties. No Throne of Denmark, but they do the job.* She hovered in front of me, explained that she did her 'cooking' on them, her word for the casting of spells.

'I live with sciatica. A deformation of the lower spine

---

* The legendary Danish Throne, in the Castle of Rosenberg, composed entirely of *Monodon Monoceros* – unicorn narwhal – tusk. Made between 1662 and 1671 by Bendix Grodtschilling. N.B. *Monodon* is no stranger to British shores: 'A specimen of this whale, measuring about eighteen feet, exclusive of the horn or tooth, was some time ago stranded on the coast of Lincolnshire, at no great distance from Boston, and was said to have been taken alive.' See George Gregory, *A Dictionary of Arts and Sciences*, Volume 2, 1816 ('Exclusive of the horn or tooth' in this instance presumably means 'already scavenged when they got there'). The species came to public attention in 2019 when Darryn Frost used a narwhal tusk (taken from its display case in Fisherman's Hall) to go after the attacker during the terrorist incident on London Bridge.

structurally incompatible with the sciatic nerve. It all centres around L7, where my spine meets the whale's. Cetacean ergonomics. These alleviate some of the pain.'

I asked if they were equivalent bones, lumbar vertebrae.

'Consecutive cranial vertebrae – what once supported a head now supports an arse.' She stirred her sage. 'It's why I do my cooking on them – so that the sorcery of any given day might be similarly steadied.'

At this moment Cat popped her head round the door; asked with a hint of frustration how much longer we were going to be.

'Half an hour?' said Aggers.

'Won't keep you much longer.'

'Apologies, it's just that we're throwing a birthday party for a friend later,' said Bones, 'A milestone. Need to do the place up.'

'Congrats. Which milestone?'

'The big nine zero zero,' said Cat.

'Nine hundred? *Many* congratulations.'

'Sixty', said Bones, looking sternly after the woman who had already left the room.

We continued chatting, eventually addressing the jaw curving upward to the ceiling.

Bones had stolen it from a manor house in the 1970s. I was in good company. A whalebone jaw had started it all off for her too. She hadn't been able to resist either, saw it on a hike one day and came back that moonlit night – didn't think it appropriate for it to serve as the entrance to private property. It had served as her 'magical cetacean gateway' ever since. A threshold between worlds – a doorway through which she might pass in order to get a better look at the one she'd just exited. For her, magic was about facilitation rather than intervention: an opening of doors.

'A kind of Heaven's Gate?' I said (vindicated).

Bones said she wasn't the first witch to have 're-appropriated' the jaws of this particular species. She asked if I knew the story of Maria Hollingsworth, the German immigrant who in 1815 constructed a shelter for herself and her daughter from a whalebone arch that stood on the banks of Oak Mere in Delamere Forest, Cheshire (the arch had been erected in the late eighteenth century).

I did. It's included by Redman.

Hollingsworth's husband had been killed in the Napoleonic Wars and she, a German by birth, had petitioned the local landlady for sanctuary: 'I vind, that all the Commons are verbid,' she wrote to one Lady Thumbley of the manor. Her arrival at the edge of the lake caused quite a stir in the local community, no doubt because she'd seemingly decided to live as a kind of unrepentant Jonah (gossips even speculated that she was Napoleon in hiding). Thumbley eventually granted permission and she lived there from 1815 to 1829.*

Bones looked at my expression; asked if anything was the matter.

Just a very familiar sensation. The dead whales at it again – returning me to a story of a German immigrant woman in England, a single mother, a child with a semi-present English father – whale jawbone. (During the divorce proceedings, my father even referred to my mother as Napoleon). I explained that she'd narrated my potted family history, and that my life

* See *The Story of the Old Woman of Delamere Forest* (Chester: J. Fletcher, printer, 1832), p. 1.

sometimes felt like a series of looping cetacean synchronicities. 'There really isn't a single thought or experience that goes unmediated. There seems to be a whale-shaped absence in my soul. Part dead whale.'

Bones looked at me, weighing my words. 'It's different for men,' she began. 'This assumed "unmediated" state of theirs. It causes them to become alarmed by anything that demonstrates their own entanglement.' Life didn't need to be a psychodrama of compromised masculinity, a melancholic identification with Leviathan's corpse; didn't have to constitute a perpetual trafficking back and forth between extremes of masculine universal floatation – and the fear of running aground; didn't need to forever rebound between fantasies of liquidity and nightmares of dependency and complicity. 'Maybe that's what comet woman was up to: bequeathing you your own *stranding* in the world.'

Perhaps I ought to stay with that strandedness awhile – stop trying to wriggle free – cut ties, continually declare my sovereign independence.

'It wasn't necessary to go looking for her. You've been bound all along.'

The sage had gone out.

I asked if they'd be willing to say a prayer or cast a spell for my family, human and non-, present, past and yet to come. Bones suggested I pass through the door. Aggers helped her manoeuvre the jawbone to the centre of the room. Bones called Cat into the room and they held it up as I ducked underneath. I didn't look back.

Aggers gave me some of the sage to take home with me. They wished Alicia the best for the coming weeks.

As I was leaving, on a side table illuminated by a

lampshade, I noticed the intimately familiar cover of Redman's *Whales' Bones*. I stopped for a moment. I couldn't quite put my finger on it – it didn't look quite right to me. Sitting in the car, about to drive home, I was moved to retrieve its counterpart from my bag: *Whales' Bones of the British Isles, Supplement 2004–2010*.

How had I not seen it before? On the cover of the original *Whales' Bones*, Redman stands underneath a whalebone arch, crucified against some railings. For the *Supplement*, exactly the same image – except the jawbones have disappeared. *Vanished*.

Presumably it was Redman's son, Ed, showing off his Photoshopping skills with a visual gag – the *Supplement* includes the bones missed from the original catalogue; it's a book of gone-missing bones.

Something gave way. A moment of clarity if not release.

This paradoxical impulse of the stranded whale to take flight. Have you heard about the painting at the Fitzwilliam Museum in Cambridge – the 1641 landscape, 'View of Scheveningen Sands' by Hendrick van Anthonissen? A restoration team discovered why the group of people figured in the image had gathered on one part of the beach. Because they were once looking at a stranded whale, stolen from view by

the painter. When placed next to the restored second image, the soiled original looks as though the stranded whale was not just painted over, but tonally dissolved into the canvas at large, an elemental particle of the world that surrounds us and flows through us.

Strandings. The moment when life began attempting and failing to crawl out of the oceans. How painfully, pitifully slow the process was. The carcasses strewn on the beaches eventually signalling the necessity to develop a technology that might avoid the recursive pattern. It took millions of years, an almost total standstill, but God eventually created legs, better fins. 'Anti-stranding devices.' Strandings. The basic emergency instructing life on earth. The foundational prompt, daring the species to imagine the next mode of rescue. It all starts on the beach. No wonder humans eventually raised the whale up from the beach and placed it among the stars, the heavenly ethereal counterpart, the constellation Cetus.

If the precondition of life in the universe is water, the next cosmic law along must be the absolute necessity of that life to become stranded; to develop an awareness of that stranded state of being. The mysterious emergence of consciousness nothing more than the painful, eventual recognition that occasionally dawns after billions upon billions of beachings. The current capitalist arrangement destroying the planet: just the latest seemingly immovable mode of stranding. The corpses, human and non-, are piled up so high you'd have thought something would have given by now. But the ultra-violent expropriation of resources (human and non-) converted into private property feels so natural to so many. Capitalist culture even attempted to hijack the topos itself: the image of the creature 'successfully' pulling itself out of the primordial soup became synonymous with the victory of particular individuals who deserved their wealth and status because they were the fittest and had struggled the hardest. What a catastrophic sleight of hand. And sleight of fin.

It's not that alternative leg technology isn't readily available (capitalist political economy rampantly accelerated the dreams and possibilities of other more equitable worlds to come); it's just that fledgling alternatives tend to be immobilised and amputated as soon as they start talking about the redistribution of wealth and the imagination-stunting evils of private property. As the waters rise around us, though – as dry land literally turns to beach – what days will come? Will that renew the collective struggle? Will that focus minds, rekindle the desire to collectively demand an alternative to these living-dead cycles of profit and loss? How long must we flounder in current form? In Paris, 1968, the revolutionary cry was '*Sous les pavés, la plage!*' – 'Under the paving stones, the beach!' The beach, the sight of new imaginings, redemptive apocalyptic futures. An ushering in of the newly stranded consciousness – a new commons listening at last to what the stranded whales have been telling us all along.

Amid the British Museum's vast holdings of imperial plunder, there's an eighth-century Anglo-Saxon box: Franks Casket. It's made from the bones of a stranded whale. A Northumberland monk carved the riddle-warning into the front panel: 'The flood cast up the fish on the mountain-cliff/ The terror-king became sad when he swam on the shingle/ Whale's bone.' The basic law: those who claim leviathanic dominion will run aground. Get picked over. As for the jaw, it's probably rotting in a garden, but for me it will always remain a mutinous and magical maypole.

In 2017, the debut of another 'London Whale'. In spite of a petition signed by 32,556 protesting the desecration of a 'monument to British heritage', the Natural History Museum replaced 'Dippy' the Dinosaur with a blue whale skeleton: 'Hope'. A juvenile female. She stranded off Wexford in 1891, and her twenty-five-metre skeleton has been part of the museum's collections since. After decades on display in the mammals' gallery, the 130-year-old specimen was restored and re-hung in Hintze Hall, posed in iron girders as if diving down from the arched ceiling to feed on a ball of krill.

> BLUE WHALE TAKES CENTRE STAGE AT NATURAL HISTORY MUSEUM. The specimen is being given the name 'Hope' as a 'symbol of humanity's power to shape a sustainable future'. Blue whales are now making a recovery following decades of exploitation that nearly drove them out of existence.
>
> Jonathan Amos, BBC News, 13 July 2017

> THE SECRET HISTORY OF HOPE THE BLUE WHALE HAS FINALLY BEEN REVEALED. New research looking at the chemical make-up of Hope's baleen plates has revealed where the blue whale travelled, and that she may have been pregnant in her last year of life.
>
> Josh Davis, Natural History Museum, 21 September 2018

CLIMATE PROTESTERS STAGE 'DIE-IN' AT NATURAL HISTORY MUSEUM AS DEMONSTRATIONS ENTER SECOND WEEK.

Telegraph, 22 April 2019

There's a photograph of me carrying Rosa, emerging from the sliding doors of the Royal Devon and Exeter Hospital. She's wrapped in a blanket of blue whales. I'm smiling, even though I've just witnessed Alicia's forty-two hours of labour: her story to tell. My absolute powerlessness and fear, the clutching of hands as the next wave of pain arrived and crested, and then the deranging beauty, eyes open, staring up through the water of the birthing pool backlit by purple – straight through and beyond me. Into their future of terrible uncertainty, but where the promise of new kinship and association must still surely beckon.

Predictably enough, Rosa's life began surrounded by

cetaceans, carved and printed and stuffed. Nearly every children's book she received from friends and relatives concerned whales, quite often stranded ones.* Alicia's observation: 'In all of these stories, the stranded whale swims back out to sea and lives happily ever after. The perpetual, recursive fantasy.'

In the first year of the pandemic we moved to County Durham – a new job, and a chance to live within striking distance of the east coast. Made the decision just after our second daughter Meri arrived, and just before everything changed. An enforced stranding no one could have anticipated. We live in a former pit village. Walk to the local church and you can see as far as the start of the North York Moors; along the eastern stretch of coastline the combined afterglow of Middlesbrough, Hartlepool and Sunderland. Their beaches of sand, microfossils and coal slag. In the middle distance, there's a recently completed Amazon warehouse the size of a major international airport.

I didn't have to wait long. Christmas Eve, just as Britain officially cut its ties with Europe, another mass stranding. Brexit bookended by mass whale death: the very beginning of 2016; the very end of 2020. Ten sperm whales between Withernsea and Tunstall, a few miles north of Spurn Point and close to Burton Constable Hall. The news quickly ebbed away. No one had the stomach for this season's greeting.

On the evening of Christmas Day, Alicia saw me fidgeting and asked when I was setting off. I said I'd be back by midday and arrived at sunrise, having driven through the

* Two copies of Donaldson and Scheffler's *The Snail and the Whale* (2003), three copies of Davies' *The Storm Whale* (2013), and one of Case's *Emma and the Whale* (2017).

night past multiple warnings on the M1 to 'Stay Local.' The peak of the second wave – six whales lying in pairs along this kilometre stretch. Unlike anything I'd ever seen, except in Hideo Kojima's *Death Strandings*, a video game released at the end of 2019. Thirty-five wind turbines immediately out to sea, their aquatic orchestral force no doubt contributing to a final distressed decision to make for land. The whales now lying slumped on this cold British beach; on this isolated landmass that had just moved heaven and earth to reaffirm a series of backward-looking ideas about nationhood, free trade and sovereignty.

Three trickles of gore, streaks of bloody lightning, ran down the soft gradient of the sand beach towards the water. The familiar sensation of interposing my own body between a whale and the sea as I hopped over. There was hardly anyone there. A dog walker greeted me with a friendly 'Merry Christmas'; a couple of teenagers making the rounds from the previous night; a classically dressed member of the necro-cetaecean subculture holding a geological hammer and pretending to look for fossils. I could see from two of the animals that he'd spent the first part of the morning knocking out teeth, four of them by my count. He was walking back home across the dunes now. With everyone still involved with Christmas these animals were ripe for the picking.

On the radio they keep talking about silence; the body released from its sonic carapace; the newly quiet skies amplifying the birdsong; the whales now singing clearly and sweetly in the oceans.

But stand next to a stranded whale and hear this other song. The song of an apocalypse that has already happened.

The song of a post-mortem world, a once-taken-for-granted future turned aftermath. Of the comet that's come and gone. The tree planting, the recycling, the tiny percentage of accumulated capital donated to charity: zombie life-support. So many buckets of water poured on Leviathan's head. So many other possible futures. Listen to the song of the dead and dying whales. The blast blasted blubber beyond all believable bounds.

Illustration credits

3 Sohail Akhter/Alamy Stock Photo; 7 José Javier Delgado Esteban/Whale Nation Studio © 2011–2021; 8 © Daryl Hinds 2016; 11 Yendis/Alamy Stock Photo; 12 Reproduced by kind permission of The British Library; 47 Still from *Werckmeister Harmonies*, directed by Béla Tarr and Ágnes Hranitzky (2001); 57 © Marina Diane Rees; 59 © Lee Post; 71 © Illustrated London News Ltd/Mary Evans; 74 © Matt Hutton, VintageLooker; 75 PA Images/Alamy Stock Photo; 84 Still from *Moby-Dick*, directed by John Huston (1956), © MGM; 87 © Illustrated London News Ltd/ Mary Evans; 88 Still from *The Monster that Challenged the World*, dir. Arnold Laven (1957), © MGM; 89 © Illustrated London News Ltd/Mary Evans; 97 Trinity Mirror/Mirrorpix/Alamy Stock Photo; 105 *Newsweek*, April 7, 1997. Photograph © Peter Riley; 106 Copyright © Peter Riley. Permission is granted to copy, distribute and/or modify this document under the terms of the GNU Free Documentation License, Version 1.2 or any later version published by the Free Software Foundation; with no Invariant Sections, no Front-Cover Texts, and no Back-Cover Texts. A copy of the license is included in the section entitled 'GNU Free Documentation License'; 118 Photograph © Peter Riley; 120

© Antiquariat Steffen Völkel GmbH; 121 Royal Collection Trust/© Her Majesty Queen Elizabeth II 2021; 122 © The British Museum Images; 132 Robert Stainforth/Alamy Stock Photo; 135 © John Cole, 'JP Morgan London Whale' (2012); 136 © RJ Matson, *St. Louis Post-Dispatch*; 137 © David Simonds and the *Observer* newspaper (2012); 144 © Alicia Williamson; 153 From *Punch's Almanack*. Campwillowlake/iStock; 159 FLHC 20212/Alamy Stock Photo; 164 © Molly Crabapple; 176 Historic Images/Alamy Stock Photo; 177 © Dundee Art Galleries and Museums/Bridgeman Images; 187 © Bonhams; 193 © Paul Davies; 198 Trinity Mirror/Mirrorpix/Alamy Stock Photo; 202–4 © Oliver Hoare Ltd; 205 *The Sun*, 18 July 2015. Photograph © Peter Riley; 220 Photograph © Peter Riley; 225 Phil Clarke Hill/Getty Images; 229 Photograph © Peter Riley.

Acknowledgements

First I need to thank the two people who saw promise in this book and made it possible. Working with Ed Lake, my editor at Profile Books, has been one of the great pleasures and privileges of my career. I will be forever indebted to the sheer force of his intelligence, patience and understanding. The insight and encouragement of Chris Wellbelove, my agent at Aitken Alexander Associates, has also been an extraordinary and inspiring gift. Thank you both for setting up the Ideas Prize, and for giving me this life-changing opportunity. I'd also like to thank everyone at Profile for seeing this book into print: Lottie Fyfe, Graham Coster, Georgia Poplett, Anna-Marie Fitzgerald and Rebecca Gray.

I am indebted to my early readers and friends, particularly Mark Steven, Merlin Sheldrake, Andy Brown, Rob Macfarlane, Elliot Kendall and Daisy Hay. Particular thanks and love to the brilliant Sam North, Benedict Morrison and Jim Blackstone for commenting so helpfully on extended drafts. As usual, the perspicacity and generosity of Tom Evans kept me honest, and helped me see the way through.

My love, admiration and gratitude to Mark Storey, Ceri Gorton, Michael Jonik, Rob Turner, Lucy Powell, Kate Montague, Henry Power, Karen Edwards, Laura Salisbury, Regenia

Gagnier, Vike Plock, Lara Choksey, Chris Campbell, Treasa De Loughry, Ellie Stedall, Steve Graves, Mariangela Milioto, Lloyd Pratt, Rodrigo Andrés, Pablo Cáceres Silguero, Tom F. Wright, Noreen Masud, Dan Grausam, Dan Hartley, Stephen Regan, Ed Sugden, Jay Grossman, Wyn Kelley, John Bryant, Sam Otter and Timothy Marr.

Given the occasionally sensitive nature of this book, I am only able to publicly acknowledge a fraction of those people who were instrumental to its composition. My sincere thanks to Daryl Hind, Philippa Wood, Marina Rees, Nicholas Redman, Lee Post, Shannon Muchow, José Javier Delgado Esteban, John Cole, Robert Matson, David Simonds, Adam Partridge, Paul Davies and Damian Hoare.

To those who remain in the woodwork: solidarity, love, respect and gratitude always.

To my father: thank you for your loyalty, support and love. So pleased to have drawn closer to you and Christine in recent years. All love to my wonderful American family: Wendy, Peter, Evan, Mel, Medora, Eos and Iona.

Alicia, I know how fortunate I am to be your partner in life. I draw strength from your courage, intellect, love and integrity every day. Thank you for all the guidance and encouragement – for calling me out when I needed calling out, and for sharing with me your wisdom, body and grace. Rosa and Meri, I love you: God only knows what you'll make of this. I'm happy to answer any questions as and when.

I'll try and avoid extrapolation now. I just want to record what happened. The deadline for these acknowledgements was 7 December 2021. Five weeks earlier, the beginning of November, I was walking down Hallgarth Street in Durham and saw something lying on the pavement. I stopped and stared down at a humpback whale, bowed in the act of marking a turn. I think it's part of a brooch – an anonymous piece of metal without hallmark (if anyone wants it back, just get in touch).

I picked it up and put it in my wallet. That evening, maybe it was early the next morning, my mum called me to say she was feeling unwell. She received her diagnosis two weeks later: stage four lung cancer with metastasis to the liver. She died on Monday 29 November at 1.15 p.m.

We managed to bring her home for her final days – two of her friends, Jim, district and Macmillan nurses, Alicia and myself keeping vigil. She could have had no doubts concerning the love and care that surrounded her. When Storm Arwen hit, ours was the only village in the vicinity not to lose power and water. I'm grateful to have held her when she died. Just she and I for that last hour. I was able to tell her that her love was my foundation.

She had wanted the windows of her room to be open. From her morphine-lorazepam half-sleep we communicated with a squeeze of hands. As she stopped breathing, at the precise moment of her passing, she opened her eyes and said something. I couldn't make out what, but I swear I heard something fly straight out of the open window.

I feel the grief. The cancer killed her so quickly. Took away a wonderful Oma, mother-in-law and friend. Rosa decorated her room; Meri serenaded her with nursery rhymes; Alicia read to her, played music for her, stepped up whenever I felt I couldn't go on. My mum felt the force of our love – long before this – and it made her happy. So many glimmers of hope and redemption in the darkness. Her love and care is our foundation. So this is dedicated to my mum, Claudia Riley, 1949–2021.